Common Core Math Workouts Grade 6

AUTHORS: Karise Mace and Keegen Gennuso
EDITORS: Mary Dieterich and Sarah M. Anderson
PROOFREADER: Margaret Brown

COPYRIGHT © 2014 Mark Twain Media, Inc.

ISBN 978-1-62223-469-1

Printing No. CD-404220

Mark Twain Media, Inc., Publishers
Distributed by Carson-Dellosa Publishing LLC

Visit us at www.carsondellosa.com

Table of Contents
With Common Core State Standard Correlations

The corresponding Common Core State Standard for Mathematics is listed at the beginning of each exercise below.

Table of Contents
With Common Core State Standard Correlations (cont.)

Introduction to the Teacher

The time has come to raise the rigor in our children's mathematical education. The Common Core State Standards were developed to help guide educators and parents on how to do this by outlining what students are expected to learn throughout each grade level. The bar has been set high, but our students are up to the challenge.

This worktext is designed to help teachers and parents meet the challenges set forth by the Common Core State Standards. It is filled with skills practice and problem-solving practice exercises that correspond to each standard for mathematics. With a little time each day, your students will become better problem solvers and will acquire the skills they need to meet the mathematical expectations for their grade level.

Each page contains two "workouts." The first workout is a skills practice exercise, and the second is geared toward applying that skill to solve a problem. These workouts make great warm-up or assessment exercises. They can be used to set the stage and teach the content covered by the standards. They can also be used to assess what students have learned after the content has been taught.

We hope that this book will help you help your students build their Common Core Math strength and become great problem solvers!

Karise Mace and Keegen Gennuso

Name: _____   Date: _____

GEOMETRY – Area

CCSS Math Content 6.G.A.1: Find the area of right triangles, other triangles, special quadrilaterals, and polygons by composing into rectangles or decomposing into triangles and other shapes; apply these techniques in the context of solving real-world mathematical problems.

SHARPEN YOUR SKILLS:

Use what you know about calculating the area of right triangles and rectangles to calculate the area of the isosceles trapezoid below.

Area = _____

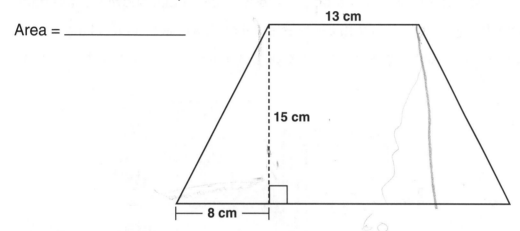

APPLY YOUR SKILLS:

Geneva is working on a quilt with her grandmother. They plan to make a queen-sized quilt, which is 60 inches by 80 inches. They will make squares using the design below and then assemble the squares to make the quilt. How much of each color material will they need for the entire quilt? (NOTE: Assume that all trapezoids shown are isosceles. You should also assume that both red trapezoids are equivalent to each other and that both blue trapezoids are equivalent to each other.)

Red material: _____

· White material: _____

Blue material: _____

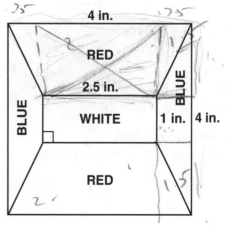

Name: _____ Date: _____

GEOMETRY – Volume

CCSS Math Content 6.G.A.2: Find the volume of a right rectangular prism with fractional edge lengths by packing it with unit cubes of the appropriate unit fraction edge lengths, and show that the volume is the same as would be found by multiplying the edge lengths of the prism. Apply the formulas to find volumes of right rectangular prisms with fractional edge lengths in the context of solving real-world mathematical problems.

SHARPEN YOUR SKILLS:

The right rectangular prism shown below left can be packed with the cubes shown on the right. Each cube is $\frac{3}{10}$ centimeter by $\frac{3}{10}$ centimeter by $\frac{3}{10}$ centimeter. Show how the product of the length, width, and height of the prism is equal to the total volume of $\frac{3}{10}$ centimeter unit cubes that can be packed into the prism.

Volume of rectangular prism:

Volume of all the unit cubes:

_____

APPLY YOUR SKILLS:

Let's Get Moving is a company that packs and moves people. One of the types of truck that they use has a trailer that is a right rectangular prism.

1. Calculate the volume of the trailer.

    _____

2. The movers typically use a box that is $2\frac{1}{2}$ feet by $2\frac{1}{2}$ feet by $2\frac{1}{2}$ feet. Calculate the volume of one of those boxes.

    _____

3. Then, calculate how many of those boxes can fit in the trailer.

    _____

Name: _____ Date: _____

GEOMETRY – Polygons

CCSS Math Content 6.G.A.3: Draw polygons in the coordinate plane given coordinates for the vertices; use coordinates to find the length of a side joining points with the same first coordinate or the same second coordinate. Apply these techniques in the context of solving real-world and mathematical problems.

SHARPEN YOUR SKILLS:

1. Plot the points given by the ordered pairs: A (–5, 6), B (4, 6), C (1, –7), D (–8, –7).

2. Connect the points in alphabetical order. Then, connect point D to point A to form a quadrilateral.

3. How long is side DC? Explain how you determined your answer.

   _____

   _____

   _____

   _____

   _____

   _____

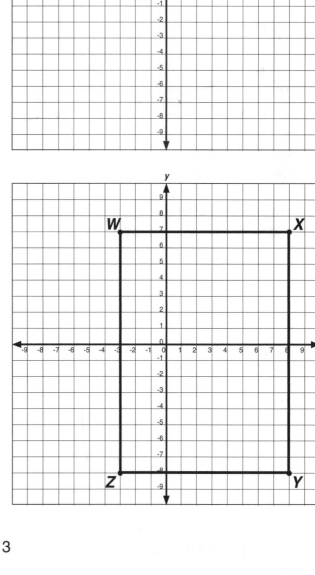

APPLY YOUR SKILLS:

1. Determine the perimeter of the figure

   WXYZ. _____

2. Explain how you determined your answer.

   _____

   _____

   _____

   _____

   _____

Name: _____ Date: _____

GEOMETRY – Solids

CCSS Math Content 6.G.A.4: Represent three-dimensional figures using nets made up of rectangles and triangles, and use the nets to find the surface area of these figures. Apply these techniques in the context of solving real-world and mathematical problems.

SHARPEN YOUR SKILLS:

Sketch and label the net of the square pyramid shown below. Then, calculate its surface area.

Surface Area = _____

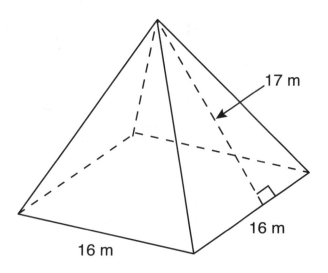

APPLY YOUR SKILLS:

Bart wants to wrap the package shown below. He has 600 square inches of wrapping paper. Does he have enough to wrap the package? Explain how you determined your answer.

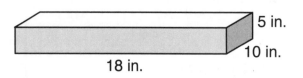

_____

_____

_____

_____

_____

Name: _____ Date: _____

RATIOS AND PROPORTIONAL RELATIONSHIPS – Ratios

CCSS Math Content 6.RP.A.1: Understand the concept of a ratio and use ratio language to describe a ratio relationship between two quantities.

SHARPEN YOUR SKILLS:

Explain what each ratio below means.

1. The ratio of apples to bananas is 4:1.

_____

2. The ratio of oranges to apples is 1:2.

_____

3. The ratio of grapes to oranges is 16:1.

_____

APPLY YOUR SKILLS:

Write a sentence describing the ratio relationship between:

1. the white marbles and the gray marbles.

_____

_____

2. the white marbles and the black marbles.

_____

_____

3. the gray marbles and the black marbles.

_____

_____

Name: _____ Date: _____

RATIOS AND PROPORTIONAL RELATIONSHIPS – Unit Rates

CCSS Math Content 6.RP.A.2: Understand the concept of a unit rate *a/b* associated with a ratio *a:b* with *b* ≠ 0, and use rate language in the context of a ratio relationship.

SHARPEN YOUR SKILLS:

Write the ratio and unit rate for each of the following.

1. We traveled 300 miles in 5 hours. _____

2. An ice maker produces 3 batches of ice in 12 hours. _____

3. Adam was paid $36 for 4 hours of work. _____

4. Evie read 32 pages in 60 minutes. _____

5. Whitney can knit 2 scarves in the same amount of time it takes Glenna to knit 3 hats.

_____

APPLY YOUR SKILLS:

Write the ratio and unit rate for each of the following. Then, explain what the unit rate means.

Ingredients for Trail Mix

3 cups peanuts

1 cup chocolate pieces

4 cups mini pretzels

2 cups raisins

5 cups sunflower seeds

3 cups dried apricots

1. Peanuts to sunflower seeds _____

_____

_____

2. Mini pretzels to raisins _____

_____

_____

3. Dried apricots to peanuts _____

_____

_____

4. Sunflower seeds to dried apricots _____

_____

_____

5. Chocolate pieces to raisins _____

_____

_____

Name: _____   Date: _____

RATIOS AND PROPORTIONAL RELATIONSHIPS –
Tables and Equivalent Ratios

CCSS Math Content 6.RP.A.3a: Make tables of equivalent ratios relating quantities with whole-number measurements, find missing values in the tables, and plot the pairs of values on the coordinate plane. Use tables to compare ratios.

SHARPEN YOUR SKILLS:

The table shows the relationship between the number of trees and the number of people for which the trees produce oxygen for a year.

1. Fill in the missing values in the table.
2. Plot the pairs of values on the coordinate plane.
3. Write a statement about the rate at which trees produce oxygen using the table and graph.

_____

_____

Number of Trees	Number of People for Which the Trees Produce Oxygen for a Year
1	2
4	8
10	
	26
18	

APPLY YOUR SKILLS:

The tables show the number of rotations two different gears in a clock make over given amounts of time. Write the ratios for the number of rotations to hours for each gear. Then on your own paper, compare these ratios and explain what you think they might indicate about the size of the gears.

Gear #1   Ratio: _____

Number of Hours	Number of Rotations
2	20
5	50
12	120
24	240

Gear #2   Ratio: _____

Number of Hours	Number of Rotations
2	120
5	300
12	720
24	1440

Name: _____ Date: _____

RATIOS AND PROPORTIONAL RELATIONSHIPS –
Unit Rate Problems

CCSS Math Content 6.RP.A.3b: Solve unit rate problems including those involving unit pricing and constant speed.

SHARPEN YOUR SKILLS:

A small automobile factory manufactures 4 cars in 3 days. At that rate, how many cars can the factory manufacture in 30 days? Explain how you determined your answer.

_____

_____

_____

_____

_____

APPLY YOUR SKILLS:

The Great Speedster roller coaster runs 3 trains every 10 minutes. At that rate, how many trains does the roller coaster run every hour? Explain how you determined your answer.

_____

_____

_____

_____

_____

Name: _____ Date: _____

RATIOS AND PROPORTIONAL RELATIONSHIPS –
Percents as Rates

CCSS Math Content 6.RP.A.3c: Find a percent of a quantity as a rate per 100 (e.g., 30% of a quantity means 30/100 times the quantity); solve problems involving finding the whole, given a part and the percent.

SHARPEN YOUR SKILLS:

A local bookstore finds that each day, approximately 20% of their sales are nonfiction books. If they sell 425 books one day, how many of those books are nonfiction? Show your work.

_____

APPLY YOUR SKILLS:

Anastasia works at the local museum and has been tracking museum attendance. On Thursday, she counts 220 museum visitors under the age of 12. She misplaces her count of the total number of visitors that day, but she knows that the under-12 visitors were 55% of the total visitors. Determine the total number of visitors to the museum on Thursday. Show your work.

_____

Name: _____ Date: _____

RATIOS AND PROPORTIONAL RELATIONSHIPS –
Ratios and Unit Conversion

CCSS Math Content 6.RP.A.3d: Use ratio reasoning to convert measurement units; manipulate and transform units appropriately when multiplying or dividing quantities.

SHARPEN YOUR SKILLS:

Robert Wadlow currently holds the record as the world's tallest man. He was 107 inches tall. Calculate Robert's height in feet and inches. Show your work. (Hint: 12 inches = 1 foot)

_____

APPLY YOUR SKILLS:

Mark is picking up a load of concrete highway dividers. Each divider weighs about 3 tons and 1,500 pounds. He can haul a load that weighs up to 72 tons. Determine the maximum number of concrete dividers Mark can haul at one time. Show your work. (Hint: 1 ton = 2000 pounds)

_____

Name: _____ Date: _____

THE NUMBER SYSTEM – Compute With Fractions

CCSS Math Content 6.NS.A.1: Interpret and compute quotients of fractions, and solve word problems involving division of fractions by fractions, e.g., by using visual fraction models and equations to represent the problem.

SHARPEN YOUR SKILLS:

Determine the quotient of $\frac{3}{5}$ and $\frac{2}{3}$.

1. Write an equation for the statement above. _____

2. Draw a model that you could use to evaluate the equation from part 1.

3. Evaluate the equation. Show your work and explain how you know your answer is correct.

_____

_____

_____

APPLY YOUR SKILLS:

Francesca's garden is $\frac{3}{4}$ acre. She would like to divide her garden into $\frac{1}{8}$ acre sections. Into how many $\frac{1}{8}$ acre sections is Francesca able to divide her garden?

1. Write an equation to model the situation above. _____

2. Draw a model that you could use to evaluate the equation from part 1.

3. Evaluate the equation. Into how many $\frac{1}{8}$ acre sections is Francesca able to divide her garden?

_____

_____

_____

Name: _____ Date: _____

THE NUMBER SYSTEM – Dividing Multi-Digit Numbers

CCSS Math Content 6.NS.B.2: Fluently divide multi-digit numbers using the standard algorithm.

SHARPEN YOUR SKILLS:

Calculate the quotient.

1. $26\overline{)4,810}$

2. $249\overline{)31,623}$

3. $354\overline{)34,695}$

APPLY YOUR SKILLS:

Mr. Shelley asks his students to determine the quotient of 82,399 divided by 214. The work of two students is shown below. Determine which student calculated the quotient correctly, and then explain what the other student did wrong.

Student #1

```
        3859
214 | 82,399
     –642
      1819
     –1712
      1079
     –1070
         9
        –9
         0
```

Student #2

```
        385 R9
214 | 82,399
     –642
      1819
     –1712
      1079
     –1070
         9
```

Right or Wrong?

_____

_____

_____

Name: _____   Date: _____

THE NUMBER SYSTEM – Operations With Decimals

CCSS Math Content 6.NS.B.3: Fluently add, subtract, multiply, and divide multi-digit decimals using the standard algorithm for each operation.

SHARPEN YOUR SKILLS:

Add or subtract as indicated.

1. 153.482
    + 46.216

3. 432.86
    + 8.942

2. 589.754
    – 279.123

4. 81.3429
    – 9.276

APPLY YOUR SKILLS:

The table below lists the weight, diameter, and thickness of some U.S. coins. Use the data in the table to answer the questions. Show your work.

	Penny	Nickel	Dime	Quarter	Half-Dollar	Presidential $1
Weight	2.5 g	5 g	2.268 g	5.67 g	11.34 g	8.1 g
Diameter	0.75 in.	0.835 in.	0.705 in.	0.955 in.	1.205 in.	1.043 in.
Thickness	1.55 mm	1.95 mm	1.35 mm	1.75 mm	2.15 mm	2 mm

1. How much heavier is a quarter than a dime?

    _____

2. If you stack one of each coin on top of the other, how tall will the stack be? _____

3. If you lay one of each coin side by side on the table, how many inches long will the row of coins be?

    _____

Dime = 0.705 in.

Half-Dollar = 1.205 in.

Name: _____ Date: _____

THE NUMBER SYSTEM – Operations With Decimals

CCSS Math Content 6.NS.B.3: Fluently add, subtract, multiply, and divide multi-digit decimals using the standard algorithm for each operation.

SHARPEN YOUR SKILLS:

Multiply and divide as indicated.

1.   5.814
     x   2.9

3.   3.2$\overline{)82.0768}$

2.   235.8
     x 61.18

4.   1.23$\overline{)531.483}$

APPLY YOUR SKILLS:

Deborah plans to cook a turkey for a family gathering. She is trying to determine how big of a turkey to purchase. Her grandmother tells her that the rule of thumb is 0.75 pound of meat for each adult in attendance. She finds a 19.83 pound turkey at the grocery store.

1. How many adults could Deborah feed with the 19.83 pound turkey? _____

2. There will be 22 adults at Deborah's family gathering. If each of them eats 0.75 pound of turkey, will the 19.83 pound turkey provide enough meat to feed them? Explain how you determined your answer.

_____

_____

3. The turkey is on sale for $1.29 per pound. How much will the turkey cost? _____

Name: _____ Date: _____

THE NUMBER SYSTEM – GCF and LCM

CCSS Math Content 6.NS.B.4: Find the greatest common factor of two whole numbers less than or equal to 100 and the least common multiple of two whole numbers less than or equal to 12. Use the distributive property to express a sum of two whole numbers 1 – 100 with a common factor as a multiple of a sum of two whole numbers with no common factor.

SHARPEN YOUR SKILLS:

Determine the greatest common factor of the given pair of numbers.

1. 24 and 36 _____

2. 52 and 91 _____

Determine the least common multiple of the given pair of numbers.

3. 2 and 9 _____

4. 4 and 6 _____

APPLY YOUR SKILLS:

A caterer is arranging 72 ham sandwiches and 48 pimento cheese sandwiches on serving trays. He would like to arrange the sandwiches so that each tray has the same number of each kind of sandwich. The number of ham sandwiches on a tray does not need to equal the number of pimento cheese sandwiches on that tray.

1. What is the maximum number of trays the caterer can use? Explain how you determined your answer. _____

_____

_____

2. If the caterer uses the maximum number of trays, how many of each type of sandwich will be on each tray? _____

3. The caterer has 6 silver trays and would like to use them for the sandwiches. Can the caterer divide the ham sandwiches and pimento cheese sandwiches evenly among the 6 trays? If so, how many of each type of sandwich will be on each tray? Explain how you determined your answers. _____

_____

_____

Name: _____   Date: _____

THE NUMBER SYSTEM – GCF and LCM

CCSS Math Content 6.NS.B.4: Find the greatest common factor of two whole numbers less than or equal to 100 and the least common multiple of two whole numbers less than or equal to 12. Use the distributive property to express a sum of two whole numbers 1 – 100 with a common factor as a multiple of a sum of two whole numbers with no common factor.

SHARPEN YOUR SKILLS:

Rewrite each sum using the distributive property by factoring out the GCF.

1. 12 + 15

_____

2. 28 + 49

_____

3. 25 + 55

_____

4. 72 + 30

_____

5. 50 + 75

_____

6. 42 + 54

_____

APPLY YOUR SKILLS:

Explain how rewriting 18 + 27 using the distributive property makes it easier to mentally calculate the sum. Support your answer with work.

_____

_____

_____

_____

_____

_____

Name: _____   Date: _____

THE NUMBER SYSTEM – Integers

CCSS Math Content 6.NS.C.5: Understand that positive and negative numbers are used together to describe quantities having opposite directions or values (e.g., temperature above/below zero, elevation above/below sea level, credits/debits, positive/negative electrical charge); use positive and negative numbers to represent quantities in real-world context, explaining the meaning of 0 in each situation.

SHARPEN YOUR SKILLS:

Represent each of the following verbal expressions with an integer.

1. Beatrice withdrew 28 dollars from her checking account. _____

2. The average temperature in Dallas, Texas, in July is 36 degrees above zero on the Celsius scale. _____

3. Pikes Peak is approximately 14,114 feet above sea level. _____

4. The lowest temperature ever recorded was approximately 129 degrees below zero on the Fahrenheit scale. _____

5. The quarterback lost 8 yards on the first play. _____

APPLY YOUR SKILLS:

Explain the meaning of 0, positive integers, and negative integers in each of the following scenarios.

1. A checking account _____

    _____

2. A Celsius thermometer _____

    _____

3. Land elevation _____

    _____

4. An elevator _____

    _____

5. Yards gained or lost during a football game _____

    _____

Name: _____ Date: _____

THE NUMBER SYSTEM –
Graphing Integers on a Number Line

CCSS Math Content 6.NS.C.6a: Recognize opposite signs of numbers as indicating locations on opposite sides of 0 on the number line; recognize that the opposite of the opposite of a number is the number itself, e.g., –(–3) = 3, and that 0 is its own opposite.

SHARPEN YOUR SKILLS:

Describe where the number is located on the number line. Then, write the number that is at the exact opposite side of 0 on the number line.

1. 4 _____
_____

2. –9 _____
_____

3. 10 _____
_____

4. 0 _____
_____

5. –3 _____
_____

6. 6 _____
_____

7. 5 _____
_____

8. –8 _____
_____

-12 -11 -10 -9 -8 -7 -6 -5 -4 -3 -2 -1 0 1 2 3 4 5 6 7 8 9 10 11 12

APPLY YOUR SKILLS:

Write the number that is described in the exercise.

1. The opposite of –3 _____

2. The opposite of 6 _____

3. The opposite of –7 _____

4. The opposite of 0 _____

5. The opposite of the opposite of 4 _____

6. The opposite of the opposite of –5 _____

7. The opposite of the opposite of the opposite of 8 _____

8. The opposite of the opposite of the opposite of –2 _____

Name: _____   Date: _____

THE NUMBER SYSTEM – Ordered Pairs

CCSS Math Content 6.NS.C.6b: Understand signs of numbers in ordered pairs as indicating locations in quadrants of the coordinate plane; recognize that when two ordered pairs differ only by signs, the locations of the points are related by reflections across one or both axes.

SHARPEN YOUR SKILLS:

Without graphing, determine the quadrant in which the point represented by the ordered pair would be graphed. Explain how you determined your answer.

1. (4, 2) _____

_____

2. (–5, –6) _____

_____

3. (7, –3) _____

_____

4. (–9, 8) _____

_____

APPLY YOUR SKILLS:

Without graphing, explain the relationship between the given points. Explain how you determined your answers.

1. *A* (–3, 5) and *B* (–3, –5) _____

_____

2. *A* (–3, 5) and *B* (3, –5) _____

_____

3. *A* (–3, 5) and *B* (3, 5) _____

_____

Name: _____   Date: _____

THE NUMBER SYSTEM –
Graphing Integers and Real Numbers

CCSS Math Content 6.NS.C.6c: Find and position integers and other rational numbers on a horizontal or vertical number line diagram; find and position pairs of integers and other rational numbers on a coordinate plane.

SHARPEN YOUR SKILLS:

Identify which point corresponds to the given integer or real number.

1. 2.5 _____

2. –6 _____

3. 9 _____

4. –5.1 _____

5. –2.5 _____

6. 6 _____

7. –9 _____

8. 5.1 _____

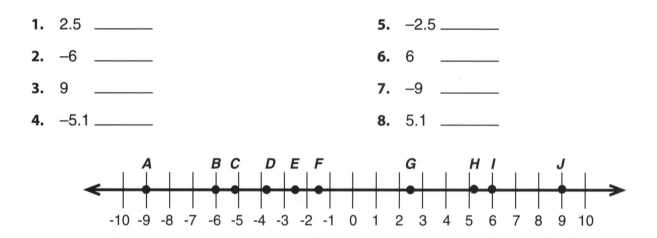

APPLY YOUR SKILLS:

Graph the integer or real number on the number line and label it with the given letter.

1. *N:* –5.5

2. *E:* 2

3. *R:* 7.9

4. *I:* –9.4

5. *G:* 0

6. *T:* –4

7. *E:* –1

8. *S:* 9.5

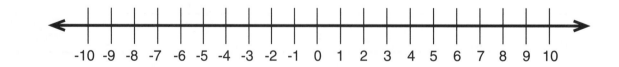

Name: _____ Date: _____

THE NUMBER SYSTEM –
Graphing Integers and Real Numbers

CCSS Math Content 6.NS.C.6c: Find and position integers and other rational numbers on a horizontal or vertical number line diagram; find and position pairs of integers and other rational numbers on a coordinate plane.

SHARPEN YOUR SKILLS:

Identify which point corresponds to the given coordinates.

1. $(3, -7)$ _____

2. $(6.5, 2)$ _____

3. $(-3, 0)$ _____

4. $(-2, 6\frac{1}{2})$ _____

5. $(-7, -3)$ _____

6. $(0, 3)$ _____

7. $(\frac{13}{2}, -2)$ _____

8. $(7, 3)$ _____

APPLY YOUR SKILLS:

Plot the point on the coordinate plane and label it with the given letter.

1. $A\,(4, 0)$

2. $B\,(0, -7)$

3. $C\,(7, -4)$

4. $D\,(-5, -5\frac{1}{10})$

5. $E\,(1.9, 9)$

6. $F\,(-\frac{15}{2}, 4.5)$

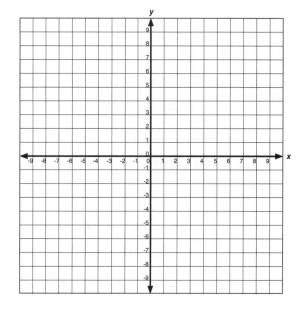

Name: _____ Date: _____

THE NUMBER SYSTEM – Interpreting Inequality Statements

CCSS Math Content 6.NS.C.7a: Interpret statements of inequality as statements about the relative position of two numbers on a number line diagram.

SHARPEN YOUR SKILLS:

Write a statement about the relationship between the locations on a number line diagram of the numbers in the given inequality.

1. $-7 < 4$ _____

2. $-5 > -8$ _____

3. $6 < 10$ _____

4. $2 > -3$ _____

5. $0 < 9$ _____

6. $0 > -1$ _____

APPLY YOUR SKILLS:

1. Carmen claims that the inequalities $-5 < 3$ and $3 > -5$ mean the same thing. Is she correct? Explain your reasoning. _____

    _____

    _____

    _____

2. Stephanie says that if $4 > 2$ is a true inequality, then $-4 > -2$ must also be a true inequality. Is she correct? Explain your reasoning. _____

    _____

    _____

    _____

3. Jacob was absent yesterday and doesn't understand how to interpret the inequality $-1 > -6$. He asks you to explain it to him. Write a short paragraph explaining it to him. _____

    _____

    _____

    _____

Name: _____   Date: _____

THE NUMBER SYSTEM – Ordering Rational Numbers

CCSS Math Content 6.NS.C.7b: Write, interpret, and explain statements of order for rational numbers in real-world contexts.

SHARPEN YOUR SKILLS:

Write an inequality to express the relationship described in the exercise.

1. −2 is less than 4   _____

2. 7 is greater than −7   _____

3. −5 is greater than −10   _____

4. 2 is less than 6   _____

5. 8 is greater than 0   _____

6. −3 is less than 0   _____

7. −9 is less than −1   _____

8. 4 is greater than −6   _____

APPLY YOUR SKILLS:

1. Janessa claims that the expression −7°F < −1°F expresses the fact that −1°F is warmer than −7°F. Bart claims that the expression −1°F > −7°F expresses the fact that −1°F is warmer than −7°F. Who is correct? Explain how you determined your answer.

    _____

    _____

    _____

2. Write an inequality to express the fact that a checking account balance of −$56 is less than a balance of $21. Explain how you determined your answer.

    _____

    _____

Name: _____ Date: _____

THE NUMBER SYSTEM – Absolute Value

CCSS Math Content 6.NS.C.7c: Understand the absolute value of a rational number as its distance from 0 on the number line; interpret absolute value as magnitude for a positive or negative quantity in a real-world situation.

SHARPEN YOUR SKILLS:

Complete the statement with < or >. Then, explain how you determined your answer.

1. $|18|$ _____ $|7|$ _____

_____

2. $|-22|$ _____ $|-14|$ _____

_____

3. $|-3|$ _____ $|8|$ _____

_____

4. $|35|$ _____ $|-42|$ _____

_____

5. $|-2.9|$ _____ $|-3.2|$ _____

_____

6. $|\frac{3}{4}|$ _____ $|-\frac{1}{4}|$ _____

_____

7. $|-\frac{2}{5}|$ _____ $|\frac{3}{7}|$ _____

_____

8. $|6.39|$ _____ $|9.63|$ _____

_____

APPLY YOUR SKILLS:

Ruth just got a new prescription for eyeglasses. The strength of a lens is measured in diopters. A negative diopter strength helps correct for nearsightedness, and a positive diopter strength helps correct for farsightedness. The strength of her left lens will be –3.25 diopters, and the strength of her right lens will be 2.00 diopters. Which eye needs more correction? Explain how you determined your answer.

_____

_____

Name: _____ Date: _____

THE NUMBER SYSTEM –
Comparing Absolute Value and Statements of Order

CCSS Math Content 6.NS.C.7d: Distinguish comparisons of absolute value from statements about order.

SHARPEN YOUR SKILLS:

1. Client #1 has a checking account balance of –$82. Client #2 has a checking account balance of –$79. Which client has a greater debt? Explain how you determined your answer.

_____

_____

_____

2. City A has an elevation of –5 feet. City B has an elevation of –8 feet. Which city is more below sea level? Explain how you determined your answer.

_____

_____

_____

3. The average temperature in January of Cold Town is –11°F. The average temperature in January of Shiver City is –7°F. Which city is colder on average during January? Explain how you determined your answer.

_____

_____

_____

APPLY YOUR SKILLS:

Explain why –8 is less than –2 but the absolute value of –8 is greater than the absolute value of –2. Use a number line diagram and/or an example to support your answer.

_____

_____

_____

_____

Name: _____   Date: _____

THE NUMBER SYSTEM – Problem Solving With Graphing

CCSS Math Content 6.NS.C.8: Solve real-world and mathematical problems by graphing points in all four quadrants of the coordinate plane. Include use of coordinates and absolute value to find distances between points with the same first coordinate or the same second coordinate.

SHARPEN YOUR SKILLS:

Create a map of AnyTown, USA by plotting the following points on the coordinate grid. Then, use your map to answer the questions. Explain how you determined your answers. (NOTE: Each grid-line represents 1 block.)

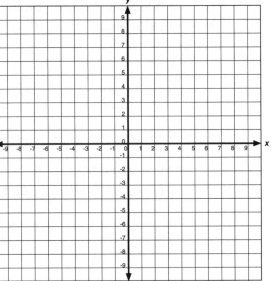

- Library (–3, 6)
- Grocery Store (–10, –9)
- High School (4, 4)
- Elementary School (0, –5)
- Town Hall (–7, 0)
- Post Office (0, 8)
- Middle School (4, –3)
- Hardware Store (6, 6)

1. How far is the library from the hardware store?

_____

_____

2. How far is the high school from the middle school?

_____

_____

3. How far is the post office from the elementary school?

_____

_____

APPLY YOUR SKILLS:

Thomas measured and recorded the temperature every couple of days for the last 10 days. He graphed his results on the grid below. Today's temperature is graphed at (0, 5). He also graphed the predicted temperatures for the coming week. Eight days ago, the temperature was –6°F. It has risen since then, but is supposed to drop again to –6°F four days from now. How many days will have elapsed between the two recorded instances of –6°F? Explain how you determined your answer.

Temperature (°F)

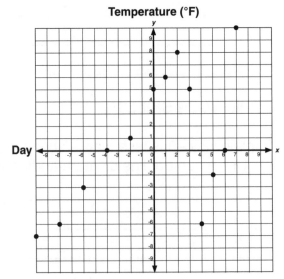

_____

_____

_____

_____

Name: _____ Date: _____

EXPRESSIONS AND EQUATIONS –
Numerical Expressions With Exponents

CCSS Math Content 6.EE.A.1: Write and evaluate numerical expressions involving whole-number exponents.

SHARPEN YOUR SKILLS:

Evaluate.

1. 4^3

2. $6^2 + 3$

3. $(2 + 7)^2$

4. $5^4 - 10^2$

5. $45 - (8^3 \div 4^2)$

6. $(2^5 - 3^3)^4 + 23$

APPLY YOUR SKILLS:

Mrs. Booth asks her students to evaluate the expression $(8 - 3)^3 - 2^5 + 4^3$. The work of two students is shown below. Which student evaluated the expression correctly? Identify and explain the mistake(s) the other student made.

Student #1
$(8 - 3)^3 - 2^5 + 4^3 = 5^3 - 2^5 + 4^3$
$= 125 - 32 + 64$
$= 93 + 64$
$= 157$

Student #2
$(8 - 3)^3 - 2^5 + 4^3 = 5^3 - 2^5 + 4^3$
$= 125 - 32 + 64$
$= 125 - 96$
$= 29$

_____

_____

_____

_____

Name: _____ Date: _____

EXPRESSIONS AND EQUATIONS – Writing Expressions

CCSS Math Content 6.EE.A.2a: Write expressions that record operations with numbers and with letters standing for numbers.

SHARPEN YOUR SKILLS:

Write the phrase as a mathematical expression.

1. Add x and 8.

    _____

2. Subtract 9 from p.

    _____

3. Calculate the difference of 7 and s.

    _____

4. Multiply 4 and t.

    _____

5. Determine the quotient of r and 12.

    _____

6. Calculate the sum of 2 and q.

    _____

7. Determine the product of y and –6.

    _____

8. Divide 10 by w.

    _____

APPLY YOUR SKILLS:

Write two different phrases for the mathematical expression.

1. $15 \div t$

    _____

    _____

2. $4 + m$

    _____

    _____

3. $d - 5$

    _____

    _____

4. $\dfrac{f}{4}$

    _____

    _____

5. $3g$

    _____

    _____

6. $x + 11$

    _____

    _____

7. $8 - b$

    _____

    _____

8. $a \times 2$

    _____

    _____

Name: _____   Date: _____

EXPRESSIONS AND EQUATIONS –
Identifying Parts of Expressions

CCSS Math Content 6.EE.A.2b: Identify parts of an expression using mathematical terms (sum, term, product, factor, quotient, coefficient); view one or more parts of an expression as a single entity.

SHARPEN YOUR SKILLS:

Identify the parts of the expression using mathematical terms.

1. $18 \div 3$

_____

_____

2. $5(6 - 2)$

_____

_____

3. $4 + (6 \div 3)$

_____

_____

_____

4. $8 + 7$

_____

_____

5. 3×9

_____

_____

6. $(10 \div 5) + (6 \times 2)$

_____

_____

_____

APPLY YOUR SKILLS:

Write an expression that satisfies the given description.

1. The sum of three terms _____

2. The product of two factors _____

3. The sum of two terms, where the second term is the difference of two terms

_____

4. The quotient of two terms _____

5. The product of two factors, where the first factor is the sum of two terms and the second factor is the quotient of two terms _____

6. The quotient of two terms, where the first term is the product of two factors and the second term is the difference of two terms _____

Name: _____　Date: _____

EXPRESSIONS AND EQUATIONS – Evaluating Expressions

CCSS Math Content 6.EE.A.2c: Evaluate expressions at specific values of their variables. Include expressions that arise from formulas used in real-world problems. Perform arithmetic operations, including those involving whole-number exponents, in the conventional order when there are no parentheses to specify a particular order (Order of Operations).

SHARPEN YOUR SKILLS:

1. Evaluate $3x + 9$, when $x = 4$.

2. Evaluate $5p^3$, when $p = -2$.

3. Evaluate $s^2 + \frac{3}{8}$, when $s = \frac{1}{2}$.

4. Evaluate $y \div 7$, when $y = -35$.

5. Evaluate $(28 - r)^5 + r$, when $r = 26$.

6. Evaluate $48 \div d + 9$, when $d = -3$.

APPLY YOUR SKILLS:

1. Use the formula $C = \frac{5}{9}(F - 32)$, where F represents degrees Fahrenheit and C represents degrees Celsius, to convert 77°F to degrees Celsius. Show your work.

_____

2. The distance, in feet, a falling object travels in a given amount of seconds can be determined by the formula $d = 16t^2$, where d is the distance in feet and t is the time in seconds. How far will an object fall in 3 seconds? Show your work.

_____

_____

Name: _____   Date: _____

EXPRESSIONS AND EQUATIONS –
Generating Equivalent Expressions

CCSS Math Content 6.EE.A.3: Apply the properties of operations to generate equivalent expressions.

SHARPEN YOUR SKILLS:

Write an equivalent expression for the given expression.

1. $x + x + x + x$

_____

2. $3(2r + 7)$

_____

3. $5y - 2y$

_____

4. $5(6 - p)$

_____

5. $-25w - 15z$

_____

6. $-3t + 4q + 5t - 7q$

_____

7. $-8(4a + 9b)$

_____

8. $8c + 72d$

_____

APPLY YOUR SKILLS:

1. Mrs. Wood asks her students to write an equivalent expression for the expression $-36x + 48y$. The answers of four students are shown below. Which student(s) wrote a correct equivalent expression? Identify and correct the mistake(s) the other students made. Explain how you determined your answer.

Student #1: $-12(3x - 4y)$ Student #3: $-3(12x + 16y)$
Student #2: $12(3x + 4y)$ Student #4: $-4(9x - 12y)$

_____

_____

_____

_____

2. A student was absent the day your teacher taught you how to write equivalent expressions. He is having trouble understanding why $16ab$ is not an equivalent expression for $2(3a + 5b)$. Write a short explanation to the student.

_____

_____

_____

Name: _____ Date: _____

EXPRESSIONS AND EQUATIONS –
Identifying Equivalent Expressions

CCSS Math Content 6.EE.A.4: Identify when two expressions are equivalent (i.e., when the two expressions name the same number regardless of which value is substituted into them.)

SHARPEN YOUR SKILLS:

Select the expression from the column on the right that is equivalent to the expression in the column on the left and write its corresponding letter in the blank.

_____	**1.**	$9x - 3x + 4x$	**A.**	$-20x + 24$
_____	**2.**	$-4(5x - 6)$	**B.**	$2x$
_____	**3.**	$-4(5x + 6)$	**C.**	$-10x$
_____	**4.**	$20x + 24x$	**D.**	$10x$
_____	**5.**	$4(5x + 6)$	**E.**	$20x + 24$
_____	**6.**	$8x + 3x - 9x$	**F.**	$44x$
_____	**7.**	$4x - 24x + 10x$	**G.**	$-20x - 24$
_____	**8.**	$4(5x - 6)$	**H.**	$20x - 24$

APPLY YOUR SKILLS:

Circle all of the expressions that are equivalent to the expression $-6(8a + 4b)$.

$-48a - 24b$ $\qquad\qquad\qquad\qquad\qquad$ $-12(4a + 2b)$

$\qquad\qquad$ $-4(12a + 6b)$ $\qquad$ $24(-2a + b)$

$\qquad\qquad\qquad\qquad$ $6(-8a - 4b)$ $\qquad\qquad$ $-2(24a + 12b)$

$\qquad\qquad$ $-6(8a - 4b)$ $\qquad\qquad$ $3(-16a + 8b)$

$24(-2a - b)$ $\qquad\qquad$ $-2a - 2b$ $\qquad\qquad\qquad$ $-12(4a - 2b)$

Name: _____   Date: _____

EXPRESSIONS AND EQUATIONS –
Solving Equations and Inequalities

CCSS Math Content 6.EE.B.5: Understand solving an equation or inequality as a process of answering a question: which values from a specified set, if any, make the equation or inequality true? Use substitution to determine whether a given number in a specified set makes an equation or inequality true.

SHARPEN YOUR SKILLS:

Use substitution to determine whether or not the given value is a solution to the given equation or inequality. Explain how you determined your answer.

1. $x = 2; 2x + 5 = 9$

_____

_____

_____

2. $x = 3; -4x = 12$

_____

_____

_____

3. $x = 45; \dfrac{x}{7} = 6$

_____

_____

_____

4. $x = -4; 8 - 10x = 48$

_____

_____

_____

5. $x = 5; 3x \geq 9$

_____

_____

_____

6. $x = -3; -4x + 7 < 19$

_____

_____

_____

7. $x = -18; \dfrac{x}{-3} > 5$

_____

_____

_____

8. $x = 6; 7 + 3x \leq 6$

_____

_____

_____

APPLY YOUR SKILLS:

Haru claims that 4, 9, and 23 are all solutions to the inequality $-5x + 7 \leq -13$. In fact, he says that any number greater than or equal to 4 is a solution to that inequality. Is Haru correct? Explain your reasoning.

_____

_____

_____

_____

Name: _____ Date: _____

EXPRESSIONS AND EQUATIONS — Writing Expressions

CCSS Math Content 6.EE.B.6: Use variables to represent numbers and write expressions when solving a real-world or mathematical problem; understand that a variable can represent an unknown number, or, depending on the purpose at hand, any number in a specified set.

SHARPEN YOUR SKILLS:

1. Adele's grandmother is 56 years older than she is. Let *a* represent Adele's age. Write an expression for Adele's grandmother's age.

   _____

2. Tyler has saved $25 less than Benjamin. Let *b* represent the amount Benjamin has saved. Write an expression for the amount Tyler has saved.

   _____

3. Cassandra has four times as many goldfish as Graham has. Let *g* represent the number of goldfish Graham has. Write an expression for the number of goldfish Cassandra has.

   _____

4. Devan, Tracy, and Scott plan to split the profits from their lemonade stand evenly. Let *p* represent their profits from the lemonade stand. Write an expression that can be used to determine how much each person receives.

   _____

APPLY YOUR SKILLS:

Caroline makes bags out of recycled juice pouches. She decides to sell the bags at a school carnival. While all of the juice pouches were donated, Caroline did have to buy thread, needles, and a few other supplies to make the bags. Her expenses totaled $23. Caroline plans to sell each bag for $5. Choose a variable to represent the number of bags Caroline sells. Then, write an expression that can be used to calculate her profit.

_____

Name: _____ Date: _____

EXPRESSIONS AND EQUATIONS –
Solving Equations in Real-World Contexts

CCSS Math Content 6.EE.B.7: Solve real-world and mathematical problems by writing and solving equations of the form and for cases in which *p*, *q*, and *x* are all nonnegative rational numbers.

SHARPEN YOUR SKILLS:

Write and solve an equation to answer the question. Show your work.

1. John runs a mile two minutes faster than his Uncle Lee. Uncle Lee runs a mile in 10 minutes. How long does it take John to run a mile?

_____

2. Restaurant A can seat three times as many customers as Restaurant B. Restaurant A can seat 129 customers. How many customers can Restaurant B seat?

_____

3. A cheetah can run twice as fast as a tiger. A cheetah can run 70 miles per hour. How fast can a tiger run?

_____

_____

4. An egg contains three more grams of protein than an avocado. An egg contains six grams of protein. How many grams of protein are in an avocado?

_____

APPLY YOUR SKILLS:

Daniel is two years older than Beatrice. Sydney is four times as old as Daniel. Sydney is 20 years old. Use equations to determine the ages of Daniel and Beatrice. Show your work.

_____

_____

Name: _____ Date: _____

EXPRESSIONS AND EQUATIONS – Writing Inequalities

CCSS Math Content 6.EE.B.8: Write an inequality of the form $x > c$ or $x < c$ to represent a constraint or condition in a real-world or mathematical problem. Recognize that inequalities of the form $x > c$ or $x < c$ have infinitely many solutions; represent solutions of such inequalities on number line diagrams.

SHARPEN YOUR SKILLS:

Graph the inequality on the number line diagram.

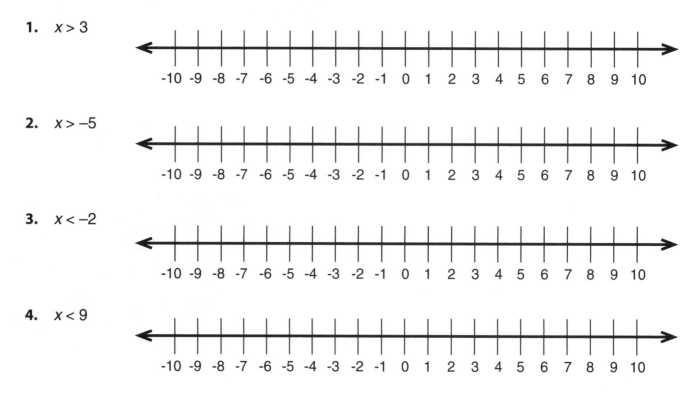

1. $x > 3$

2. $x > -5$

3. $x < -2$

4. $x < 9$

APPLY YOUR SKILLS:

Write the inequality that is represented on the number line diagram below. Then, explain what the inequality means.

_____

_____

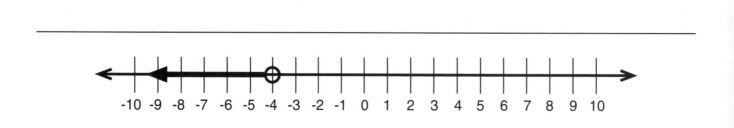

Name: _____ Date: _____

EXPRESSIONS AND EQUATIONS – Writing Inequalities

CCSS Math Content 6.EE.B.8: Write an inequality of the form $x > c$ or $x < c$ to represent a constraint or condition in a real-world or mathematical problem. Recognize that inequalities of the form $x > c$ or $x < c$ have infinitely many solutions; represent solutions of such inequalities on number line diagrams.

SHARPEN YOUR SKILLS:

Write an inequality to represent the given scenario.

1. The bank account has a balance of less than –$3. _____

2. Dominque has more than 8 bottles of nail polish. _____

3. Jared passed for more than 75 yards in the first quarter.

    _____

4. The temperature was less than 18°F. _____

5. The hikers were at an elevation of less than 45 feet above sea level. _____

6. The budget has a balance of more than –$15. _____

APPLY YOUR SKILLS:

Deisha and Barbara have each started their own babysitting business. Deisha charges a base rate of $4 per hour and then charges an additional $2 per child per hour. Barbara charges a flat rate of $10 per hour. Deisha claims that she charges less than Barbara. This situation can be represented by the inequality $2c + 4 < 10$, where c represents the number of children and $2c + 4$ represents Deisha's hourly rate. Write and graph an inequality that represents the solution set. Is Deisha's claim correct? Explain how you determined your answer.

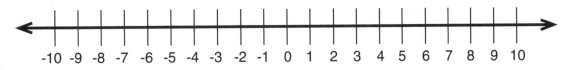

_____

_____

_____

_____

Name: _____ Date: _____

EXPRESSIONS AND EQUATIONS – Relationships Between Independent and Dependent Variables

CCSS Math Content 6.EE.C.9: Use variables to represent two quantities in a real-world problem that change in relationship to one another; write an equation to express one quantity, thought of as the dependent variable, in terms of the other quantity, thought of as the independent variable. Analyze the relationship between the dependent and independent variables using graphs and tables, and relate these to the equation.

SHARPEN YOUR SKILLS:

The Environment Club is selling reusable cloth sandwich bags to raise money for the school garden. They are selling the bags for $5 each. They spent $50 on supplies to make the bags.

1. Write an equation that gives the Environment Club's profit in terms of the number of bags sold. _____

2. Complete the table, and then graph the data.

Number of Bags	Profit (in dollars)
0	
1	
2	
5	
10	
20	

APPLY YOUR SKILLS:

Analyze your equation, table, and graph above.

1. Write a few sentences about the relationship between the equation, table, and graph.

    _____

    _____

    _____

    _____

2. How many bags must the Environment Club sell to break even? That is, when will their profit be $0? Explain how you determined your answer. _____

    _____

    _____

Name: _____ Date: _____

STATISTICS AND PROBABILITY –
Recognizing Statistical Questions

CCSS Math Content 6.SP.A.1: Recognize a statistical question as one that anticipates variability in the data related to the question and accounts for it in the answers.

SHARPEN YOUR SKILLS:

Mrs. Alexander would like to learn more about her 6th grade students. She prepares a survey for each student that includes the following questions. Explain why each question, when posed to the entire class, is or is not a statistical question.

1. How many students are in this class? _____
   _____

2. How many brothers and/or sisters do you have? _____
   _____

3. What is your favorite color? _____
   _____

4. Who is the tallest person in this class? _____
   _____

5. How many pets do you have? _____
   _____

6. How tall are you? _____
   _____

APPLY YOUR SKILLS:

1. Write three original statistical questions that you could ask students in your school. Do not use the questions from above. Explain how you know your questions are statistical.

   _____

   _____

   _____

   _____

2. Give two examples of non-statistical questions that you could ask students in your school. Do not use the questions from above. Explain how you know your questions are not statistical.

   _____

   _____

   _____

Name: _____ Date: _____

STATISTICS AND PROBABILITY – Describing Data Sets

CCSS Math Content 6.SP.A.2: Understand that a set of data collected to answer a statistical question has a distribution that can be described by its center, spread, and overall shape.

SHARPEN YOUR SKILLS:

1. Explain what it means to describe the distribution of a set of data by its center.

_____

_____

_____

_____

2. Explain what it means to describe the distribution of a set of data by its spread.

_____

_____

_____

_____

3. Explain what it means to describe the distribution of a set of data by its overall shape.

_____

_____

_____

_____

APPLY YOUR SKILLS:

Cameron says that the distribution of a data set is best described by its center. Lee, however, argues that it is best described by its spread, while Rosa claims that it is best described by its overall shape. Do you agree with Cameron, Lee, or Rosa? Explain your reasoning.

_____

_____

_____

_____

Name: _____ Date: _____

STATISTICS AND PROBABILITY –
Measures of Center and Variation

CCSS Math Content 6.SP.A.3: Recognize that a measure of center for a numerical data set summarizes all of its values with a single number, while a measure of variation describes how its values vary with a single number.

SHARPEN YOUR SKILLS:

1. Name two measures of center and explain how they are calculated.

_____

_____

_____

_____

_____

2. Name two measures of variation and explain how they are calculated.

_____

_____

_____

_____

_____

APPLY YOUR SKILLS:

Measures of center and variation are typically used together to describe a data set and its distribution. Explain which measures of center and variation should be used together and why.

_____

_____

_____

_____

_____

Name: _____   Date: _____

STATISTICS AND PROBABILITY – Displaying Data

CCSS Math Content 6.SP.B.4: Display numerical data in plots on a number line, including dot plots, histograms, and box plots.

SHARPEN YOUR SKILLS:

The table below lists the amount of vitamin C found in various foods. Create a dot plot of the data on a number line.

Food	Amount of Vitamin C (mg)
Kiwifruit	90
Brussels Sprouts	80
Strawberry	60
Pineapple	48
Grapefruit	30
Tangerine	30
Lime	30
Blackberry	21
Apricot	10
Plum	10

Food	Amount of Vitamin C (mg)
Broccoli	90
Papaya	60
Orange	53
Cantaloupe	40
Raspberry	30
Spinach	30
Mango	28
Grape	10
Watermelon	10
Avocado	8

⟵——————————————————————————⟶

APPLY YOUR SKILLS:

1. Write a statistical question that you can ask your classmates.

   _____

   _____

2. Ask your classmates the question you wrote in Exercise 1 and record the data.

3. On your own paper or on the back of this page, create a dot plot of the data you collected in Exercise 2.

Name: _____ Date: _____

STATISTICS AND PROBABILITY – Displaying Data

CCSS Math Content 6.SP.B.4: Display numerical data in plots on a number line, including dot plots, histograms, and box plots.

SHARPEN YOUR SKILLS:

The frequency table below shows the ages of U.S. Senators. Create a histogram of the data.

Age Range (in years)	Number of Senators
30 – 39	1
40 – 49	12
50 – 59	30
60 – 69	35
70 – 79	21
80 – 89	1

APPLY YOUR SKILLS:

1. Write a statistical question that you can ask your classmates.

    _____

    _____

2. Ask your classmates the question you wrote in Exercise 1 and record the data.

3. On your own paper or on the back of this page, create a histogram of the data you collected in Exercise 2.

Name: _____ Date: _____

STATISTICS AND PROBABILITY – Displaying Data

CCSS Math Content 6.SP.B.4: Display numerical data in plots on a number line, including dot plots, histograms, and box plots.

SHARPEN YOUR SKILLS:

The table shows the highest drop of 11 roller coasters. Create a box plot of the data.

Name of Roller Coaster	Drop (in feet)
Kingda Ka	418
Top Thrill Dragster	400
Steel Dragon 2000	306
Millennium Force	300
Intimidator 305	300
Goliath	255

Name of Roller Coaster	Drop (in feet)
Titan	255
Fujiyama	230
Phantom's Revenge	228
Desperado	225
Bizarro	221

APPLY YOUR SKILLS:

1. Write a statistical question that you can ask your classmates.

    _____

    _____

2. Ask your classmates the question you wrote in Exercise 1 and record the data.

3. On your own paper or on the back of this page, create a box plot of the data you collected in Exercise 2.

Name: _____ Date: _____

STATISTICS AND PROBABILITY –
Reporting the Number of Observations

CCSS Math Content 6.SP.B.5a: Summarize numerical data sets in relation to their context by reporting the number of observations.

SHARPEN YOUR SKILLS:

Determine the number of observations in the data set. Explain how you determined your answer.

1.

Range of Scores	Frequency
0 – 19	6
20 – 39	15
40 – 59	26
60 – 79	48
80 – 99	17

2.

Number of Pets	Frequency
0	6
1	5
2	3
3	1
4	12

3.

Number of Cousins	Frequency
0 – 4	14
5 – 9	11
10 – 14	14
15 – 19	21
20 – 24	5

_____   _____   _____

_____   _____   _____

_____   _____   _____

_____   _____   _____

_____   _____   _____

_____   _____   _____

_____   _____   _____

APPLY YOUR SKILLS:

The table in Exercise 2 above displays data that Mr. Sisco collected about his students and the number of pets they own. Use that data to answer the following questions.

1. How many students are in Mr. Sisco's class? Explain how you determined your answer.

_____

_____

_____

2. What is the total number of pets that the students in Mr. Sisco's class own? Explain how you determined your answer.

_____

_____

_____

_____

Name: _____ Date: _____

STATISTICS AND PROBABILITY –
Describing Attributes Under Investigation

CCSS Math Content 6.SP.B.5b: Summarize numerical data sets in relation to their context by describing the nature of the attribute under investigation, including how it was measured and its units of measurement.

SHARPEN YOUR SKILLS:

Identify the attribute under investigation. In other words, what is being determined in each situation? Then, describe how it can be measured and identify the units of measurement.

1. Maeko wants to know how many different types of fresh fruits her grocery store stocks.

    _____

    _____

    _____

2. Bob wants to determine the number of cars that pass through a traffic light between 4 p.m. and 5 p.m. in the afternoon.

    _____

    _____

3. Teresa wants to determine the rate at which her baby is growing.

    _____

    _____

    _____

APPLY YOUR SKILLS:

Demographics are measureable statistics about a population. DeJuan would like to learn more about the demographics of his school and is making a plan to collect that data.

1. List three measurable attributes of middle-school students that DeJuan could collect data about. _____

    _____

2. Explain what units DeJuan would use to measure these attributes.

    _____

    _____

3. Describe how DeJuan would collect the data for each of these attributes.

    _____

    _____

    _____

Name: _____ Date: _____

STATISTICS AND PROBABILITY –
Calculating Measures of Center and Variability

CCSS Math Content 6.SP.B.5c: Summarize numerical data sets in relation to their context by giving quantitative measures of center (median and/or mean) and variability (interquartile range and/or mean absolute deviation), as well as describing any overall pattern and any striking deviations from the overall pattern with reference to the context in which the data were gathered.

SHARPEN YOUR SKILLS:

The table shows the number of gold medals the United States has won at each Summer Olympics.

Games	Gold Medal Count
1896 Athens	11
1900 Paris	19
1904 St. Louis	78
1908 London	23
1912 Stockholm	25
1920 Antwerp	41
1924 Paris	45
1928 Amsterdam	22
1932 Los Angeles	41
1936 Berlin	24
1948 London	38
1952 Helsinki	40
1956 Melbourne	32

Games	Gold Medal Count
1960 Rome	34
1964 Tokyo	36
1968 Mexico City	45
1972 Munich	33
1976 Montreal	34
1984 Los Angeles	83
1988 Seoul	36
1992 Barcelona	37
1996 Atlanta	44
2000 Sydney	37
2004 Athens	36
2008 Beijing	36
2012 London	46

1. Calculate the median number of gold medals. Show your work.

_____

2. Calculate the interquartile range of the number of gold medals. Show your work.

_____

APPLY YOUR SKILLS:

On your own paper, write a short paragraph analyzing the U.S. Summer Olympic gold medal counts. Use the data and your calculations from above. Be sure that your analysis includes:
- An explanation of what the median means in terms of the given context
- An explanation of what the interquartile range means in terms of the given context
- A description of the overall pattern of the data
- A description of any striking deviations from the pattern and what they might mean in terms of the given context

Name: _____ Date: _____

STATISTICS AND PROBABILITY –
Calculating Measures of Center and Variability

CCSS Math Content 6.SP.B.5c: Summarize numerical data sets in relation to their context by giving quantitative measures of center (median and/or mean) and variability (interquartile range and/or mean absolute deviation), as well as describing any overall pattern and any striking deviations from the overall pattern with reference to the context in which the data were gathered.

SHARPEN YOUR SKILLS:

The table shows the number of hot dogs eaten by the winner for the last 16 years.

Year	Number of Hot Dogs	Year	Number of Hot Dogs
2012	68	2004	53.5
2011	62	2003	44.5
2010	54	2002	50.5
2009	68	2001	50
2008	59	2000	20.125
2007	66	1999	20.25
2006	53.75	1998	19
2005	49	1997	24.5

1. Calculate the mean number of hot dogs eaten by the winners. Round to the nearest tenth. Show your work.

_____

_____

2. Calculate the mean absolute deviation for the number of hot dogs eaten by the winners. Round to the nearest tenth. Show your work on your own paper.

_____

APPLY YOUR SKILLS:

On your own paper, write a short paragraph analyzing the hot dog eating results. Use the data and your calculations from above. Be sure that your analysis includes:
- An explanation of what the mean means in terms of the given context
- An explanation of what the mean absolute deviation means in terms of the given context
- A description of the overall pattern of the data
- A description of any striking deviations from the pattern and what they might mean in terms of the given context

Name: _____ Date: _____

STATISTICS AND PROBABILITY –
Relating Statistics to Shape of Distribution and Context

CCSS Math Content 6.SP.B.5d: Relating the choice of measures of center and variability to the shape of the data distribution and the context in which the data were gathered.

SHARPEN YOUR SKILLS:

Choose which measure of center (**mean** or **median**) and which measure of variability (**mean absolute deviation** or **interquartile range**) would be the best statistics to represent the data as described. Explain how you determined your answer.

1. A symmetrical data set _____

   _____

2. A data set that is skewed to the left _____

   _____

3. A data set that is symmetrical but very spread out _____

   _____

4. A data set that is skewed to the right _____

   _____

5. A data set that is mostly symmetrical but has a few extreme values _____

   _____

APPLY YOUR SKILLS:

1. Describe the shape of distribution for which the mean is the best measure of center. Then, explain why this is the case.

   _____

   _____

   _____

2. Describe the shape of distribution for which the interquartile range is the best measure of variability. Then, explain why this is the case.

   _____

   _____

   _____

Name: _____ Date: _____

STATISTICS AND PROBABILITY –
Relating Statistics to Shape of Distribution and Context

CCSS Math Content 6.SP.B.5d: Relating the choice of measures of center and variability to the shape of the data distribution and the context in which the data were gathered.

SHARPEN YOUR SKILLS:

During the last census, data was collected about household incomes. Some of the data is displayed in the histograms below. Identify which measure of center and variability you would use to describe the data that is shown in the graph. Explain your reasoning.

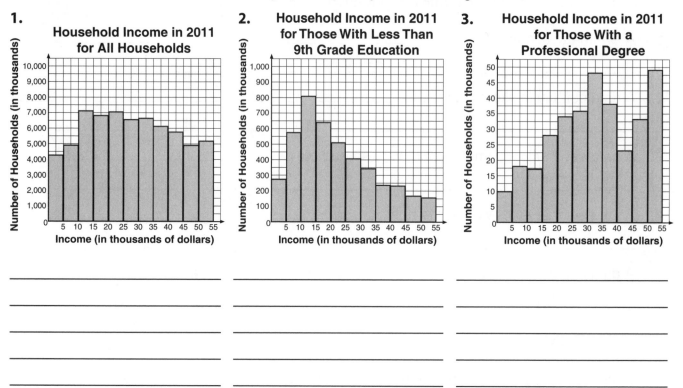

_____   _____   _____

_____   _____   _____

_____   _____   _____

_____   _____   _____

_____   _____   _____

APPLY YOUR SKILLS:

Jana has gotten the following grades on her math quizzes: 75, 82, 83, 84, 84, 87, and 88.

1. What measure of center and variability would you use to describe Jana's math quiz data? Explain your reasoning.

    _____

    _____

2. If Jana gets a 35 on her next quiz, what measure of center and variability would you use to describe Jana's math quiz data? Explain your reasoning.

    _____

    _____

ANSWER KEYS

GEOMETRY

Area (pg. 1)

SHARPEN YOUR SKILLS:

The trapezoid can be divided into 2 congruent right triangles and a rectangle.

Area of each triangle: *Area of rectangle:*

$A = \frac{1}{2}bh$ $A = bh$

$A = \frac{1}{2}(8)(15) = 60$ cm² $A = (13)(15) = 195$ cm²

The total area is 315 cm².

APPLY YOUR SKILLS:

Area of the quilt: *Area of one square:*

$A = bh$ $A = bh$

$A = 60 \times 80 = 4,800$ in.² $A = 4 \times 4 = 16$ in.²

To determine the number of squares needed, divide the total area by the area of each square; 300 squares.

Red Material:

$A = \frac{1}{2}(b_1 + b_2)h$ $A = \frac{1}{2}(4 + 2.5)(1.5)$ or 4.875 in.²

Geneva will need 2 × 4.875 or 9.75 in.² of material for each square. For the entire quilt, she will need 300 × 9.75 or 2,925 in.² of red material.

Blue Material:

$A = \frac{1}{2}(b_1 + b_2)h$ $A = \frac{1}{2}(4 + 1)(0.75)$ or 1.875 in.²

Geneva will need 2 × 1.875 or 3.75 in.² of blue material for each square. For the entire quilt, she will need 300 × 3.75 or 1,125 in.² of blue material.

White Material:

$A = bh$ $A = 2.5 \times 1$ or 2.5 in.²

Geneva will need 300 × 2.5 or 750 in.² of white material for the entire quilt.

Volume (pg. 2)

SHARPEN YOUR SKILLS:

Rectangular Prism: *Unit Cube:*

$V = l \times w \times h$ $V = l \times w \times h$

$V = \frac{9}{10} \times \frac{3}{5} \times \frac{6}{5}$ $V = \frac{3}{10} \times \frac{3}{10} \times \frac{3}{10}$

$V = \frac{162}{250}$ or $\frac{81}{125}$ cm³ $V = \frac{27}{1000}$ cm³

The volume of the rectangular prism is $\frac{81}{125}$ cubic centimeter. The volume of all of the unit cubes that can fit into the rectangular prism is $24 \times \frac{27}{1000}$ or $\frac{81}{125}$ cm³.

APPLY YOUR SKILLS:

1. $V = l \times w \times h$

     $V = 15 \times 7\frac{1}{2} \times 7\frac{1}{2}$ or $843\frac{3}{4}$ ft³

     The trailer has a volume of $843\frac{3}{4}$ cubic feet.

2. $V = l \times w \times h$

     $V = 2\frac{1}{2} \times 2\frac{1}{2} \times 2\frac{1}{2}$ or $15\frac{5}{8}$ ft³

     The volume of each box is $15\frac{5}{8}$ cubic feet.

     $843\frac{3}{4} \div 15\frac{5}{8} = \frac{3375}{4} \div \frac{125}{8} = \frac{3375}{4} \times \frac{8}{125} = \frac{27000}{500}$ or 54

     Therefore, 54 boxes will fit in the trailer.

Polygons (pg. 3)

SHARPEN YOUR SKILLS:

1. and 2.

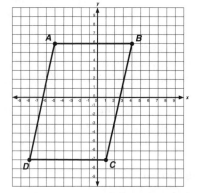

3. Points *D* and *C* have the same *y*-coordinates. So, I can calculate the absolute value of the difference between the *x*-coordinates to determine the length of side *DC*. The length of *DC* is |–8 – 1| or 9 units.

APPLY YOUR SKILLS:

Because points *W* and *X* have the same *y*-coordinates, I can calculate the absolute value of the difference between the *x*-coordinates to determine the length of *WX*. I can follow a similar procedure to calculate the length of *ZY*. Because points *W* and *Z* have the same *x*-coordinates, I can calculate the absolute value of the difference between the *y*-coordinates to determine the length of *WZ*. I can follow a similar procedure to calculate the length of *XY*.

     Length of *WX*: |–3 – 8| = 11

     Length of *ZY*: |–3 – 8| = 11

     Length of *WZ*: |7 – (–8)| = 15

     Length of *XY*: |7 – (–8)| = 15

To determine the perimeter of figure *WXYZ*, I must add up the lengths of all of the sides. Therefore, the perimeter of figure *WXYZ* is 11 + 11 + 15 + 15 or 52 units.

Solids (pg. 4)

SHARPEN YOUR SKILLS:

Square pyramid net:

Area of square base: *Area of triangular sides:*

$A = s^2$ $A = 4(\frac{1}{2}bh)$

$A = 16^2$ or 256 m² $A = 4\left(\frac{1}{2}(16)(17)\right)$ or 544 m²

The surface area is 256 + 544 or 800 m².

APPLY YOUR SKILLS:

$SA = 2lw + 2lh + 2wh$

$SA = 2(18)(10) + 2(18)(5) + 2(10)(5)$

$SA = 360 + 180 + 100$ or 640 in.²

Bart needs at least 640 in.² of wrapping paper to wrap the package. So, he does not have enough wrapping paper.

RATIOS AND PROPORTIONAL RELATIONSHIPS
Ratios (pg. 5)
SHARPEN YOUR SKILLS:
1. For every 4 apples, there is 1 banana.
2. For every 1 orange, there are 2 apples.
3. For every 16 grapes, there is 1 orange.

APPLY YOUR SKILLS:
1. The ratio of white to gray marbles is 1:4, because for every white marble, there are 4 gray marbles.
2. The ratio of white to black marbles is 1:1, because for every white marble there is 1 black marble.
3. The ratio of gray to black marbles is 4:1, because for every 4 gray marbles there is 1 black marble.

Unit Rates (pg. 6)
SHARPEN YOUR SKILLS:

1. Ratio: 300:5; Unit rate: 60 miles per hour
2. Ratio: 3:12; Unit rate: $\frac{3}{12}$ or $\frac{1}{4}$ batch of ice per hour
3. Ratio: 36:4; Unit rate: $9 per hour
4. Ratio: 32:60; Unit rate: $\frac{32}{60}$ or $\frac{8}{15}$ of a page per minute
5. Ratio: 2:3; Unit rate: $\frac{2}{3}$ of a scarf per hat

APPLY YOUR SKILLS:

1. Ratio: 3:5; Unit rate: $\frac{3}{5}$ cup of peanuts to 1 cup of sunflower seeds. For every cup of sunflower seeds in the trail mix, there is $\frac{3}{5}$ cup of peanuts.
2. Ratio: 4:2; Unit rate: 2 cups of mini pretzels to 1 cup of raisins. For every cup of raisins in the trail mix, there are 2 cups of mini pretzels.
3. Ratio: 3:3; Unit rate: 1 cup of dried apricots to 1 cup of peanuts. For every cup of peanuts in the trail mix, there is 1 cup of dried apricots.
4. Ratio: 5:3; Unit rate: $\frac{5}{3}$ or $1\frac{2}{3}$ cups of sunflower seeds to 1 cup of dried apricots. For every cup of dried apricots in the trail mix, there are $\frac{5}{3}$ or $1\frac{2}{3}$ cups of sunflower seeds.
5. Ratio: 1:2; Unit rate: $\frac{1}{2}$ cup of chocolate pieces to 1 cup of raisins. For every cup of raisins in the trail mix, there is $\frac{1}{2}$ cup of chocolate pieces

Tables and Equivalent Ratios (pg. 7)
SHARPEN YOUR SKILLS:
1.

Number of Trees	Number of People for Which the Trees Produce Oxygen for a Year
1	2
4	8
10	20
13	26
18	36

2.

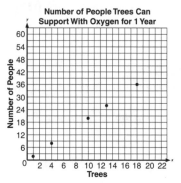

3. Answers will vary. Sample answer: Trees produce oxygen at a consistent rate. Each year, one tree produces enough oxygen for two people.

APPLY YOUR SKILLS:
For Gear #1, the ratio of the number of rotations to hours is 10:1. For Gear #2, the ratio of the number of rotations to hours is 60:1. Gear #1 has a smaller ratio than Gear #2. So, Gear #1 is bigger than Gear #2, as it rotates fewer times in 1 hour.

Unit Rate Problems (pg. 8)
SHARPEN YOUR SKILLS:
If the factory manufactures 4 cars every 3 days, then it is producing at a rate of $\frac{4}{3}$ cars per day. Therefore, in the course of 30 days, the factory manufactures $\frac{4}{3} \times 30$ or 40 cars.

APPLY YOUR SKILLS:
I know that there are 60 minutes in an hour. This means that there are six 10-minute intervals in an hour. If the Great Speedster roller coaster runs 3 trains every 10 minutes, then it will run 6×3 or 18 trains every hour.

Percents as Rates (pg. 9)
SHARPEN YOUR SKILLS:
$$\frac{20}{100} \times 425 = \frac{20}{100} \times \frac{425}{1}$$
$$= \frac{8500}{100} \text{ or } 85$$
Eighty-five of the books sold are nonfiction.

APPLY YOUR SKILLS:
$$\frac{220}{x} = \frac{55}{100}$$
$$(220)(100) = 55x$$
$$\frac{22,000}{55} = \frac{55x}{55}$$
$$400 = x$$
There were a total of 400 visitors on Thursday.

Ratios and Unit Conversion (pg. 10)
SHARPEN YOUR SKILLS:
$$107 \text{ in.} \times \frac{1 \text{ ft}}{12 \text{ in.}} = \frac{107 \text{ in.}}{1} \times \frac{1 \text{ ft}}{12 \text{ in.}}$$
$$= \frac{107 \text{ in.} \times 1 \text{ ft}}{1 \times 12 \text{ in.}}$$
$$= \frac{107 \text{ ft}}{12} \text{ or } 8 \text{ ft } 11 \text{ in.}$$

Robert Wadlow was 8 feet and 11 inches tall.

APPLY YOUR SKILLS:

Weight of One Concreter Divider:

$$3 \text{ tons} \times \frac{2000 \text{ lbs.}}{1 \text{ ton}} = \frac{3 \text{ tons}}{1} \times \frac{2000 \text{ lbs.}}{1 \text{ ton}}$$

$$= \frac{3 \text{ tons} \times 2000 \text{ lbs.}}{1 \times 1 \text{ ton}}$$

$$= \frac{6000 \text{ lbs.}}{1}$$

The weight of 1 concrete divider is 3 tons and 1,500 pounds, which is 6,000 + 1,500 or 7,500 pounds.

Weight Capacity of Mark's Trailer:

$$72 \text{ tons} \times \frac{2000 \text{ lbs.}}{1 \text{ ton}} = \frac{72 \text{ tons}}{1} \times \frac{2000 \text{ lbs.}}{1 \text{ ton}}$$

$$= \frac{72 \text{ tons} \times 2000 \text{ lbs.}}{1 \times 1 \text{ ton}}$$

$$= \frac{144,000 \text{ lbs.}}{1}$$

The weight capacity of Mark's trailer is 144,000 pounds.

Maximum Number of Concrete Dividers Mark Can Haul:

$$144,000 \text{ lbs.} \times \frac{1 \text{ divider}}{7500 \text{ lbs.}} = \frac{144,000 \text{ lbs.}}{1} \times \frac{1 \text{ divider}}{7500 \text{ lbs.}}$$

$$= \frac{144,000 \text{ lbs.} \times 1 \text{ divider}}{1 \times 7500 \text{ lbs.}}$$

$$= \frac{144,000 \text{ dividers}}{7500}$$

or 19.2 dividers

Mark can haul a maximum of 19 dividers.

THE NUMBER SYSTEM
Compute With Fractions (pg. 11)
SHARPEN YOUR SKILLS:

1. $\frac{3}{5} \div \frac{2}{3}$

2.

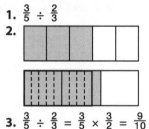

3. $\frac{3}{5} \div \frac{2}{3} = \frac{3}{5} \times \frac{3}{2} = \frac{9}{10}$

I know that the quotient of $\frac{3}{5}$ and $\frac{2}{3}$ is $\frac{9}{10}$ because $\frac{2}{3}$ of $\frac{9}{10}$ or $\frac{2}{3} \times \frac{9}{10}$ is $\frac{3}{5}$.

APPLY YOUR SKILLS:

1. $\frac{3}{4} \div \frac{1}{8}$

2.

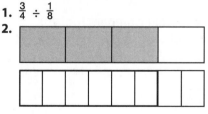

3. $\frac{3}{4} \div \frac{1}{8} = \frac{3}{4} \times \frac{8}{1} = \frac{24}{4} = 6$; Francesca can divide her garden into six $\frac{1}{8}$-acre sections.

Dividing Multi-Digit Numbers (pg. 12)
SHARPEN YOUR SKILLS:

1.
```
        185
  26 ) 4810
      −26
      221
      −208
       130
      −130
         0
```

2.
```
          127
 249 ) 31,623
      −249
       672
      −498
      1743
     −1743
         0
```

3.
```
           98 R3
 354 ) 34,695
      −3186
       2835
      −2832
          3
```

APPLY YOUR SKILLS:

Student #2 calculated the quotient correctly. Student #1 wrote the remainder as part of the whole number in the quotient. I know that this is incorrect, because 3859 × 214 is 825,826 *not* 82,399.

Operations With Decimals (pg. 13)
SHARPEN YOUR SKILLS:

1.
```
   153.482
 +  46.216
   199.698
```

2.
```
   589.754
 − 279.123
   310.631
```

3.
```
      1 1 1
    432.860
 +    8.942
    441.802
```

4.
```
            13
       7 11 2 8 12
      8 1 . 3 4 2 9
 −      9 . 2 7 6 0
      7 2 . 0 6 6 9
```

APPLY YOUR SKILLS:

1. A quarter is 3.402 grams heavier than a dime.
2. The stack would be 10.75 mm tall.
3. The row of coins would be 5.493 inches long.

Operations With Decimals (pg. 14)
SHARPEN YOUR SKILLS:

1. 16.8606
2. 14,426.244
3. 25.649
4. 432.1

APPLY YOUR SKILLS:

1. Deborah can feed 26 adults with a 19.83 lb. turkey.
2. The turkey will provide enough meat for 26 adults, so there will be more than enough meat for the 22 adults attending Deborah's family gathering.
3. The turkey will cost $25.58.

GCF and LCM (pg. 15)
SHARPEN YOUR SKILLS:

1. Factors of 24: 1, 2, 3, 4, 6, 8, 12, 24
   Factors of 36: 1, 2, 3, 4, 6, 9, 12, 18, 36
   The GCF of 24 and 36 is 12.
2. Factors of 52: 1, 2, 4, 13, 26, 52
   Factors of 91: 1, 7, 13, 91
   The GCF of 52 and 91 is 13.
3. Multiples of 2: 2, 4, 6, 8, 10, 12, 14, 16, 18,...
   Multiples of 9: 9, 18, 27, 36,...
   The LCM of 2 and 9 is 18.
4. Multiples of 4: 4, 8, 12, 16, 20,...
   Multiples of 6: 6, 12, 18, 24, 30,...
   The LCM of 4 and 6 is 12.

APPLY YOUR SKILLS:

1. Because the caterer would like the same number of ham sandwiches and the same number of pimento cheese sandwiches on each tray, I know that I need to determine the GCF of 72 and 48. The factors of 72 are 1, 2, 3, 4, 6, 8, 9, 12, 18, 24, 36, and 72. The factors of 48 are 1, 2, 3, 4, 6, 8, 12, 16, 24, and 48. The GCF of 72 and 48 is 24. Therefore, the maximum number of trays is 24.

2. If the caterer divides the sandwiches among 24 trays, there will be 3 ham sandwiches and 2 pimento cheese sandwiches on each tray.

3. Because 6 is a factor of both 72 and 48, the caterer can divide the sandwiches evenly among the 6 trays. There will be 12 ham sandwiches and 8 pimento cheese sandwiches on each tray.

GCF and LCM (pg. 16)
SHARPEN YOUR SKILLS:

1. $12 + 15 = 3(4 + 5)$
2. $28 + 49 = 7(4 + 7)$
3. $25 + 55 = 5(5 + 11)$
4. $72 + 30 = 6(12 + 5)$
5. $50 + 75 = 25(2 + 3)$
6. $42 + 54 = 6(7 + 9)$

APPLY YOUR SKILLS:

Sample answer: To rewrite $18 + 27$ using the distributive property, I must factor out the GCF of 9. The result is $9(2 + 3)$. The sum of 2 and 3 is 5. So, this expression can be simplified to $9(5)$, which is 45. It is easier to calculate the product of 9 and 5 mentally than it is to add 18 and 27, because regrouping would be necessary to calculate that sum.

Integers (pg. 17)
SHARPEN YOUR SKILLS:

1. $-\$28$
2. $36°C$
3. $14,114$ ft
4. $-129°F$
5. -8 yd

APPLY YOUR SKILLS:

1. Sample answer: In the checking account scenario, 0 represents a balance of $0. That is, the checking account does not have any money in it. Positive integers represent a positive account balance. That is, the checking account has money in it. Negative integers represent a negative account balance. That is, a person has written checks that total more than the amount of money in the checking account and therefore owes money to the bank.

2. Sample answer: In the Celsius thermometer scenario, 0 represents the temperature at which water freezes. Positive integers represent temperatures that are warmer than freezing. Negative integers represent temperatures that are colder than freezing.

3. Sample answer: In the land elevation scenario, 0 represents sea level. Positive integers represent elevations above sea level. Negative integers represent elevations below sea level.

4. Sample answer: In the elevator scenario, 0 represents the ground. Positive integers represent floors above ground. While negative integers are not typically used, they would represent floors below ground.

5. Sample answer: In the football game scenario, 0 represents the line of scrimmage. Positive integers represent a gain of yards on the play. Negative integers represent a loss of yards on the play.

Graphing Integers on a Number Line (pg. 18)
SHARPEN YOUR SKILLS:

1. 4 units to the right of zero; -4
2. 9 units to the left of zero; 9
3. 10 units to the right of zero; -10
4. At zero; 0
5. 3 units to the left of zero; 3
6. 6 units to the right of zero; -6
7. 5 units to the right of zero; -5
8. 8 units to the left of zero; 8

APPLY YOUR SKILLS:

1. 3
2. -6
3. 7
4. 0
5. 4
6. -5
7. -8
8. 2

Ordered Pairs (pg. 19)
SHARPEN YOUR SKILLS:

1. The point represented by (4, 2) will be graphed in Quadrant I, because both coordinates are positive.

2. The point represented by (−5, −6) will be graphed in Quadrant III, because both coordinates are negative.

3. The point represented by (7, −3) will be graphed in Quadrant IV, because the x-coordinate is positive and the y-coordinate is negative.

4. The point represented by (−9, 8) will be graphed in Quadrant II, because the x-coordinate is negative and the y-coordinate is positive.

APPLY YOUR SKILLS:

1. Point B is point A reflected over the x-axis. I know this because the points have the same x-coordinate, but their y-coordinates are opposites.

2. Point B is point A reflected over both the x- and y-axes. I know this because the x-coordinates are opposites, and the y-coordinates are opposites.

3. Point B is point A reflected over the y-axis. I know this because the x-coordinates are opposites, and the y-coordinates are the same.

Graphing Integers and Real Numbers (pg. 20)
SHARPEN YOUR SKILLS:

1. G
2. B
3. J
4. C
5. E
6. I
7. A
8. H

APPLY YOUR SKILLS:

Graphing Integers and Real Numbers (pg. 21)
SHARPEN YOUR SKILLS:

1. L
2. G
3. R
4. D
5. P
6. S
7. J
8. H

APPLY YOUR SKILLS:

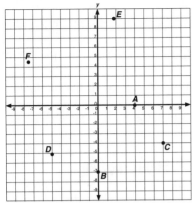

Interpreting Inequality Statements (pg. 22)
SHARPEN YOUR SKILLS:

1. −7 is to the left of 4 on a number line diagram.
2. −5 is to the right of −8 on a number line diagram.
3. 6 is to the left of 10 on a number line diagram.
4. 2 is to the right of −3 on a number line diagram.
5. 0 is to the left of 9 on a number line diagram.
6. 0 is to the right of −1 on a number line diagram.

APPLY YOUR SKILLS:

1. Carmen is correct. The inequality −5 < 3 means that −5 is to the left of 3 on a number line diagram. The inequality 3 > −5 means that 3 is to the right of −5 on a number line diagram. The statements do mean the same thing.
2. Stephanie is *not* correct. Four is to the right of 2 on a number line, but −4 is to the left of −2 on a number line.
3. The inequality −1 > −6 means that −1 is to the right of −6 on a number line. When writing negative integers on a number line, you start at zero and then move to the left. So, −1 is closer to 0 than −6 is.

Ordering Rational Numbers (pg. 23)
SHARPEN YOUR SKILLS:

1. −2 < 4   2. 7 > −7   3. −5 > −10
4. 2 < 6   5. 8 > 0   6. −3 < 0
7. −9 < −1   8. 4 > −6

APPLY YOUR SKILLS:

1. Janessa and Bart are both correct. The inequality −7°F < −1°F indicates that −7°F is less than −1°F, and the inequality −1°F > −7°F indicates that −1°F is greater than −7°F. Therefore, both inequalities indicate that −1°F is warmer than −7°F.
2. −$56 < $21; I know that −56 is less than 21. So, I can use the less than symbol to indicate that a balance of −$56 is less than a balance of $21.

Absolute Value (pg. 24)
SHARPEN YOUR SKILLS:

1. >; I know that |18| = 18, |7| = 7, and 18 > 7. So, |18| > |7|.
2. >; I know that |−22| = 22, |−14| = 14, and 22 > 14. So, |−22| > |−14|.
3. <; I know that |−3| = 3, |8| = 8, and 3 < 8. So, |−3| < |8|.

4. <; I know that |35| = 35, |−42| = 42, and 35 < 42. So, |35| < |−42|.
5. <; I know that |−2.9| = 2.9, |−3.2| = 3.2, and 2.9 < 3.2. So, |−2.9| < |−3.2|.
6. >; I know that $|\frac{3}{4}| = \frac{3}{4}$, $|-\frac{1}{4}| = \frac{1}{4}$, and $\frac{3}{4} > \frac{1}{4}$. So, $|\frac{3}{4}| > |-\frac{1}{4}|$.
7. <; I know that $|-\frac{2}{5}| = \frac{2}{5}$ or $\frac{14}{35}$, $|\frac{3}{7}| = \frac{3}{7}$ or $\frac{15}{35}$, and $\frac{14}{35} < \frac{15}{35}$. So, $|-\frac{2}{5}| < |\frac{3}{7}|$.
8. <; I know that |6.39| = 6.39, |9.63| = 9.63, and 6.39 < 9.63. So, |6.39| < |9.63|.

APPLY YOUR SKILLS:
The left eye needs more correction, because |−3.25| is greater than |2.00|.

Comparing Absolute Value and Statements of Order (pg. 25)
SHARPEN YOUR SKILLS:

1. Although −82 < −79, |−82| > |−79|, so Client #1 has a greater debt.
2. Although −5 > −8, |−5| < |−8|, so City B is more below sea level than City A.
3. Although −11 < −7, |−11| > |−7|, so Cold Town is colder than Shiver City on average during January.

APPLY YOUR SKILLS:
Because −8 is to the left of −2 on the number line, I know that −8 < −2. The absolute value of a number measures that number's distance from zero on a number line. I know that |−8| > |−2|, because −8 is further from 0 on the number line than −2 is.

Another way to think about it is in terms of temperature. A temperature of −8°F is lower than −2°F. However, −8°F is colder than −2°F.

Problem-Solving With Graphing (pg. 26)
SHARPEN YOUR SKILLS:

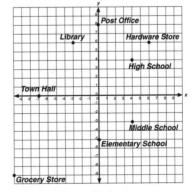

1. 9 blocks; Because the *y*-coordinates of the points representing the library and the hardware store are the same, I can add the absolute values of the *x*-coordinates to determine the distance between them. Therefore, the distance between the library and the hardware store is |−3| + |6| or 9 blocks.

2. 7 blocks; Because the *x*-coordinates of the points representing the high school and the middle school are the same, I can add the absolute values of the *y*-coordinates to determine the distance between them. Therefore, the distance between the high school and the middle school is $|4| + |–3|$ or 7 blocks.

3. 13 blocks; Because the *x*-coordinates of the points representing the post office and the elementary school are the same, I can add the absolute values of the *y*-coordinates to determine the distance between them. Therefore, the distance between the post office and the elementary school is $|8| + |–5|$ or 13 blocks.

APPLY YOUR SKILLS:

12 days; Because the *y*-coordinates of the points representing the first time the temperature was –6°F and the second time the temperature will be –6°F are the same, I can add the absolute values of the *x*-coordinates to determine the number of days between them. Therefore, the number of days that will have elapsed between the two recorded instances of –6°F is $|–8| + |4|$ or 12 days.

EXPRESSIONS AND EQUATIONS

Numerical Expressions With Exponents (pg. 27)
SHARPEN YOUR SKILLS:

1. $4^3 = 4 \times 4 \times 4$
   $= 64$

2. $6^2 + 3 = 36 + 3$
   $= 39$

3. $(2 + 7)^2 = 9^2$
   $= 81$

4. $5^4 – 10^2 = 625 – 100$
   $= 525$

5. $45 – (8^3 \div 4^2) = 45 – (512 \div 16)$
   $= 45 – 32$
   $= 13$

6. $(2^5 – 3^3)^4 + 23 = (32 – 27)^4 + 23$
   $= 5^4 + 23$
   $= 648$

APPLY YOUR SKILLS:

Student #1 evaluated the expression correctly. Student #2 added 32 and 64 before subtracting from 125.
$(8 – 3)^3 – 2^5 + 4^3 = 5^3 – 2^5 + 4^3$
$= 125 – \boxed{32 + 64}$
$= 125 – 96$
$= 29$

Writing Expressions (pg. 28)
SHARPEN YOUR SKILLS:

1. $x + 8$

2. $p – 9$

3. $7 – s$

4. $4 \times t$ or $4t$

5. $r \div 12$ or $\dfrac{r}{12}$

6. $2 + q$

7. $y \times –6$ or $–6y$

8. $10 \div w$ or $\dfrac{10}{w}$

APPLY YOUR SKILLS:

1. Sample answers: Divide 15 by *t*. Determine the quotient of 15 and *t*.

2. Sample answers: Add 4 and *m*. Calculate the sum of 4 and *m*.

3. Sample answers: Subtract 5 from *d*. Calculate the difference of *d* and 5.

4. Sample answers: Divide *f* by 4. Determine the quotient of *f* and 4.

5. Sample answers: Multiply 3 by *g*. Determine the product of 3 and *g*.

6. Sample answers: Add *x* and 11. Calculate the sum of *x* and 11.

7. Sample answers: Subtract *b* from 8. Calculate the difference of 8 and *b*.

8. Sample answers: Multiply *a* by 2. Determine the product of *a* and 2.

Identifying Parts of Expressions (pg. 29)
SHARPEN YOUR SKILLS:

1. The expression $18 \div 3$ is the quotient of the terms 18 and 3.

2. The expression $5(6 – 2)$ is the product of the factors 5 and $(6 – 2)$. The second factor is the difference of the terms 6 and 2.

3. The expression $4 + (6 \div 3)$ is the sum of the terms 4 and $6 \div 3$. The second term is the quotient of the terms 6 and 3.

4. The expression $8 + 7$ is the sum of the terms 8 and 7.

5. The expression 3×9 is the product of the factors 3 and 9.

6. The expression $(10 \div 5) + (6 \times 2)$ is the sum of the terms $10 \div 5$ and 6×2. The first term is the quotient of the terms 10 and 5. The second term is the product of the factors 6 and 2.

APPLY YOUR SKILLS:

Answers will vary for these problems. Therefore, sample answers are provided using variables to demonstrate the general form of each of the answers. Student answers should be written with numerical terms.

1. $a + b + c$

2. $a \times b$ or ab

3. $a + (b – c)$

4. $a \div b$ or $\dfrac{a}{b}$

5. $(a + b) \times (c \div d)$

6. $(a \times b) \div (c – d)$

Evaluating Expressions (pg. 30)
SHARPEN YOUR SKILLS:

1. $3x + 9 = 3(4) + 9$
   $= 12 + 9$
   $= 21$

2. $5p^3 = 5(–2)^3$
   $= 5(–8)$
   $= –40$

3. $s^2 + \dfrac{3}{8} = (\tfrac{1}{2})^2 + \dfrac{3}{8}$
   $= \dfrac{1}{4} + \dfrac{3}{8}$
   $= \dfrac{2}{8} + \dfrac{3}{8}$
   $= \dfrac{5}{8}$

4. $y \div 7 = –35 \div 7$
   $= –5$

5. $(28 – r)^5 + r = (28 – 26)^5 + 26$
   $= (2)^5 + 26$
   $= 32 + 26$
   $= 58$

6. $48 \div d + 9 = 48 \div (–3) + 9$
   $= –16 + 9$
   $= –7$

APPLY YOUR SKILLS:

1. $C = \frac{5}{9}(F - 32)$
　$= \frac{5}{9}(77 - 32)$
　$= 25$
　The temperature
　is 25°C.

2. $d = 16t^2$
　$= 16(3)^2$
　$= 144$
　The object will fall
　144 feet.

Generating Equivalent Expressions (pg. 31)
SHARPEN YOUR SKILLS:

1. $x + x + x + x = 4x$
2. $3(2r + 7) = 6r + 21$
3. $5y - 2y = 3y$
4. $5(6 - p) = 30 - 5p$
5. $-25w - 15z = -5(5w + 3z)$ or $5(-5w - 3z)$
6. $-3t + 4q + 5t - 7q = 2t - 3q$
7. $-8(4a + 9b) = -32a - 72b$
8. $8c + 72d = 8(c + 9d)$ or $4(2c + 18d)$ or $2(4c + 36d)$

APPLY YOUR SKILLS:

1. Students #1 and #4 wrote correct equivalent expressions, because they correctly factored out a common factor of both terms. Student #2 made a sign error. If you factor 12 out of both terms, then the first term will be $-3x$. Student #3 also made a sign error. If you factor -3 out of both terms, then the second term will be $-16y$.

2. Sample answer: You can use the distributive property to produce the equivalent expression of $6a + 10b$. However, you cannot combine those terms because their variables are different. So, they are not like terms. Therefore, $16ab$ is not an equivalent expression for $2(3a + 5b)$.

Identifying Equivalent Expressions (pg. 32)
SHARPEN YOUR SKILLS:

1. D　　2. A　　3. G　　4. F
5. E　　6. B　　7. C　　8. H

APPLY YOUR SKILLS:

These expressions should be circled:
$-48a - 24b$　　　$-12(4a + 2b)$
$-4(12a + 6b)$　　$6(-8a - 4b)$
$-2(24a + 12b)$　　$24(-2a - b)$

Solving Equations and Inequalities (pg. 33)
SHARPEN YOUR SKILLS:

1. Two is a solution because $2(2) + 5 = 4 + 5$ or 9.
2. Three is *not* a solution because $-4(3) = -12$.
3. Forty-five is *not* a solution, because $\frac{45}{7} = 6\frac{3}{7}$.
4. Negative 4 is a solution, because $8 - 10(-4) = 8 + 40$ or 48.
5. Five is a solution, because $3(5) = 15$, which is greater than 9.
6. Negative 3 is *not* a solution, because $-4(-3) + 7 = 12 + 7$ or 19, which is equal to 19.
7. Negative 18 is a solution, because $\frac{-18}{-3} = 6$, which is greater than 5.

8. Six is not a solution, because $7 + 3(6) = 7 + 18$ or 25, which is greater than 6.

APPLY YOUR SKILLS:

Haru is correct. If we substitute any number greater than or equal to 4 into the inequality, the inequality will be true. For example, let $x = 4$. Then, $-5(4) + 7 = -20 + 7$ or -13, which is equal to -13. Therefore, we can say that 4 is a solution to the inequality. Similarly, let $x = 9$. Then, $-5(9) + 7 = -45 + 7$ or -38, which is less than -13. Therefore, we can say that 9 is also a solution to the inequality. Finally, let $x = 23$. Then, $-5(23) + 7 = -115 + 7$ or -108, which is less than -13. Therefore, we can also say that 23 is a solution to the inequality.

Writing Expressions (pg. 34)
SHARPEN YOUR SKILLS:

1. $a + 56$　　　　2. $b - 25$
3. $4g$　　　　　4. $p \div 3$ or $\frac{p}{3}$

APPLY YOUR SKILLS:

Sample answer: Let b represent the number of bags Caroline sells. Then, her profit can be determined using the expression $5b - 23$.

Solving Equations in Real-World Contexts (pg. 35)
SHARPEN YOUR SKILLS:

1. Let m represent the time it takes for John to run a mile. Then, $m + 2 = 10$. Therefore, John runs a mile in 8 minutes.
2. Let b represent the number of people Restaurant B can seat. Then, $3b = 129$. Therefore, Restaurant B can seat 43 people.
3. Let t represent the speed at which a tiger can run. Then, $2t = 70$. Therefore, a tiger can run 35 miles per hour.
4. Let a represent the amount of protein an avocado contains. Then, $a + 3 = 6$. Therefore, an avocado contains 3 grams of protein.

APPLY YOUR SKILLS:

Let d represent Daniel's age and b represent Beatrice's age. Then, $20 = 4d$. Therefore, Daniel is 5 years old.
I know that Daniel is two years older than Beatrice, so $b + 2 = 5$. Therefore, Beatrice is 3 years old.

Writing Inequalities (pg. 36)
SHARPEN YOUR SKILLS:

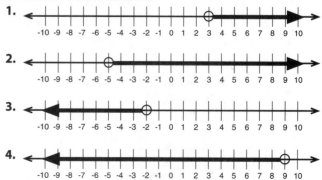

APPLY YOUR SKILLS:
The number line represents the inequality $x < -4$. So, any number that is less than −4 will satisfy the inequality.

Writing Inequalities (pg. 37)
SHARPEN YOUR SKILLS:

1. $x < -3$ **2.** $x > 8$ **3.** $x > 75$
4. $x < 18$ **5.** $x < 45$ **6.** $x > -15$

APPLY YOUR SKILLS:

$$2c + 4 < 10$$
$$2c < 6$$
$$c < 3$$

←————————————————————○————————————→
-10 -9 -8 -7 -6 -5 -4 -3 -2 -1 0 1 2 3 4 5 6 7 8 9 10

Deisha is partially correct. When substituting any number less than 3 for c, the inequality is true. For example, let $x = 2$. Then, $2(2) + 4 = 4 + 4$ or 8, and 8 is less than 10. Therefore, 2 is one solution to the inequality $2c + 4 < 10$. However, if there are 3 children in the family, Deisha's rate is $10, which is equal to Barbara's rate. If there are more than 3 children in the family, Deisha's rate is higher than Barbara's. Further, although a negative number would satisfy the inequality, it does not make sense to have a negative number of children. Therefore, Deisha does charge less than Barbara only if she is babysitting 1 or 2 children.

Relationships Between Independent and Dependent Variables (pg. 38)
SHARPEN YOUR SKILLS:

1. Let p represent profit, and let b represent the number of bags sold. Then, $p = 5b - 50$.

2.

Number of Bags (in dollars)	Profit
b	$p = 5b - 50$
0	−50
1	−45
2	−40
5	−25
10	0
20	50

Environment Club's Sandwich Bag Sale

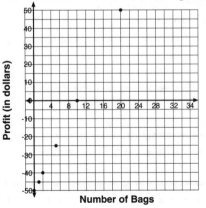

Number of Bags

APPLY YOUR SKILLS:

1. Answers will vary. However, the responses should indicate that the student understands that as the number of bags sold increases, the profit also increases. Further, students should recognize that as the number of bags sold increases by 1, the profit increases by 5, which is the coefficient of b in the equation.

2. The Environment Club must sell 10 bags to break even. I looked in my table and identified the number of bags that would produce a profit of $0.

STATISTICS AND PROBABILITY
Recognizing Statistical Questions (pg. 39)
SHARPEN YOUR SKILLS:
The questions in exercises 1 and 4 are not statistical, because there is only one answer.
The questions in exercises 2, 3, 5, and 6 are statistical, because there is variability in the answers.

APPLY YOUR SKILLS:

1. Answers will vary. However, the questions should allow for variability in the answers.

2. Answers will vary. However, the questions should *not* allow for variability in the answers.

Describing Data Sets (pg. 40)
SHARPEN YOUR SKILLS:

1. To describe the distribution of a data set by its center, you are giving one "average" value as a summary of the distribution of the data.

2. To describe the distribution of a data set by its spread, you are describing how much variation there is in the data from the least value to the greatest value.

3. To describe the distribution of a data set by its overall shape, you are describing where the highs and lows are and giving an idea about how much of the data is in those areas.

APPLY YOUR SKILLS:
Answers may vary. However, students should realize that it is most effective to describe the distribution of a data set by its center, spread, and overall shape.

Measures of Center and Variation (pg. 41)
SHARPEN YOUR SKILLS:

1. Mean and median; The mean is determined by calculating the sum of all of the data values and then dividing by the number of data values in the data set. The median is determined by listing the data values in order from least to greatest and then identifying the number that is in the middle.

2. Mean absolute deviation and interquartile range; To determine the mean absolute deviation, first calculate the absolute values of the difference between each data value and the mean of the data set. Then, calculate the mean of those values. To determine the interquartile range, first determine the median of both the lower fifty percent of the data (quartile 1) and the upper fifty percent of the data (quartile 3). Then, calculate the difference between those values.

APPLY YOUR SKILLS:

The mean and mean absolute deviation should be used together, because the mean absolute deviation gives the average of how much each data value in the set varies from the mean. The median and interquartile range should be used together, because the interquartile range gives the spread of the middle fifty percent of the data set.

Displaying Data (pg. 42)
SHARPEN YOUR SKILLS:

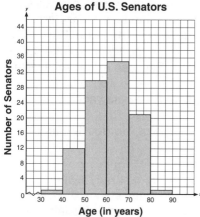

Amount of Vitamin C in Various Fruits and Vegetables

Amount of Vitamin C (mg)

APPLY YOUR SKILLS:

Answers will vary.

Displaying Data (pg. 43)
SHARPEN YOUR SKILLS:

Ages of U.S. Senators

Number of Senators

Age (in years)

APPLY YOUR SKILLS:

Answers will vary.

Displaying Data (pg. 44)
SHARPEN YOUR SKILLS:

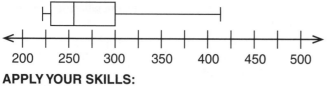

APPLY YOUR SKILLS:

Answers will vary.

Reporting the Number of Observations (pg. 45)
SHARPEN YOUR SKILLS:

1. There are 112 observations. I totaled the numbers in the frequency column to determine the number of observations.
2. There are 27 observations. I totaled the numbers in the frequency column to determine the number of observations.
3. There are 65 observations. I totaled the numbers in the frequency column to determine the number of observations.

APPLY YOUR SKILLS:

1. There are 27 students in Mr. Sisco's class. I totaled the numbers in the frequency column to determine my answer.
2. The students in Mr. Sisco's class own a total of 62 pets. To calculate this total, I first multiplied the number of pets in each row by the number of students who own that number of pets. Then I found the sum of those products. $(0)(6) + (1)(5) + (2)(3) + (3)(1) + (4)(12) = 0 + 5 + 6 + 3 + 48 = 62$

Describing Attributes Under Investigation (pg. 46)
SHARPEN YOUR SKILLS:

1. Sample answer: Maeko needs to determine how many different fruits her grocery store stocks. She can do this by counting the different kinds of fruit she sees on display. She will measure this in "types of fruit."
2. Sample answer: Bob needs to determine the number of cars that pass through the light between 4 p.m. and 5 p.m. He can do this by counting the cars that pass through the light between 4 p.m. and 5 p.m. for several days and then calculate an average. He will measure this in "cars."
3. Sample answer: Teresa needs to determine the rate at which her baby is growing. She can do this by weighing her baby each week to determine how much weight the baby is gaining. She will measure this in "pounds."

APPLY YOUR SKILLS:

1. Sample answers: weight, height, age
2. Sample answers: pounds, feet and inches, years
3. Sample answer: DeJuan could ask students to complete a survey that included their weight, height, and age. Although it would be less efficient, he could acquire a scale and tape measure and record the weight, height, and age of each student.

Calculating Measures of Center and Variability (pg. 47)
SHARPEN YOUR SKILLS:

1. The data values in order are: 11, 19, 22, 23, 24, 25, 32, 33, 34, 34, 36, 36, 36, 36, 37, 37, 38, 40, 41, 41, 44, 45, 45, 46, 78, 83. Because there is an even number of data values, the median is the average of the middle values.

   So, the median is $\frac{36 + 36}{2} = \frac{72}{2}$ or 36.

2. Q1: 32   Q3: 41

   IQR = Q3 – Q1 = 9   The interquartile range is 9.

APPLY YOUR SKILLS:

Sample answer: The data is slightly skewed to the left. The interquartile range is 9, which means the data is not very spread out. That is, for the middle fifty percent of the data, there is only a difference of 9 medals. So, for half of the years the United States has participated in the Summer Olympics, they have won between 32 and 41 gold medals. The median is 36, which means that fifty percent of the Summer Olympic gold medal counts are 36 or less. There were three Summer Olympics in which the gold medal count did not seem to follow the overall pattern of the rest of the data. In 1896, the U.S. won only 11 gold medals. In 1904, the U.S. won 78 gold medals, and in 1984, the U.S. won 83 gold medals.

Calculating Measures of Center and Variability (pg. 48)
SHARPEN YOUR SKILLS:
1. Add all the numbers in the Number (of Hot Dogs) column, divide the sum by 16, and round to the nearest tenth. The mean is about 47.6 hot dogs.

2.

Year	Number (of Hot Dogs)	Difference from Mean	Absolute Difference from Mean
2012	68	68 – 47.6 = 20.4	20.4
2011	62	62 – 47.6 = 14.4	14.4
2010	54	54 – 47.6 = 6.4	6.4
2009	68	68 – 47.6 = 20.4	20.4
2008	59	59 – 47.6 = 11.4	11.4
2007	66	66 – 47.6 = 18.4	18.4
2006	53.75	53.75 – 47.6 = 6.15	6.15
2005	49	49 – 47.6 = 1.4	1.4
2004	53.5	53.5 – 47.6 = 5.9	5.9
2003	44.5	44.5 – 47.6 = –3.1	3.1
2002	50.5	50.5 – 47.6 = 2.9	2.9
2001	50	50 – 47.6 = 2.4	2.4
2000	20.125	20.125 – 47.6 = –27.475	27.475
1999	20.25	20.25 – 47.6 = –27.35	27.35
1998	19	19 – 47.6 = –28.6	28.6
1997	24.5	24.5 – 47.6 = –23.1	23.1

Add all the numbers in the Absolute Difference from Mean column, divide the sum by 16, and round to the nearest tenth. 219.775 ÷ 16 ≈ 13.7
The mean absolute deviation is about 13.7 hot dogs.
APPLY YOUR SKILLS:
Sample answer: The hot dog eating data is slightly skewed to the left. The mean number of hot dogs eaten by the winners is about 47.6 hot dogs. That means that, on average, the hot dog eating winners ate about 47.6 hot dogs. The mean absolute deviation is about 13.7 hot dogs. That means that, on the average, the amount of hot dogs the winner ate each year varies from the mean number of hot dogs by about 13.7 hot dogs. That is a relatively large mean absolute deviation, which indicates the data is fairly spread out.

Relating Statistics to Shape of Distribution and Context (pg. 49)
SHARPEN YOUR SKILLS:
1. The mean and mean absolute deviation are the best statistics to use for symmetrical data that has no extreme values.
2. The median and interquartile range are the best statistics to use for skewed data.
3. The mean and mean absolute deviation are the best statistics to use for symmetrical data that has no extreme values.
4. The median and interquartile range are the best statistics to use for skewed data.
5. The median and interquartile range are the best statistics to use for data that has extreme values or outliers.
APPLY YOUR SKILLS:
1. The mean is the best measure of center for a symmetrical distribution. It is strongly influenced by outliers or extreme data points. So, it is better to use it when the data does not have as much variability or skew.
2. The interquartile range is the best measure of variability for data that is skewed or has outliers or extreme data points. It gives the spread of the middle fifty percent of the data. Because extreme values are not in this middle fifty percent, it gives a better idea of how the middle portion of the data varies.

Relating Statistics to Shape of Distribution and Context (pg. 50)
SHARPEN YOUR SKILLS:
1. Because this histogram is relatively symmetrical, I would use the mean and mean standard deviation.
2. Because the histogram is skewed to the left, I would use the median and interquartile range.
3. Because the histogram is skewed to the right, I would use the median and interquartile range.
APPLY YOUR SKILLS:
1. The data is relatively symmetrical. So, I would use the mean and mean absolute deviation to describe the data.
2. A quiz score of 35 is an extreme value and skews the data. So, I would use the median and interquartile range to describe the data.

3 1170 00956 1659